Point processes with a generalized order statistic property

vom Fachbereich Mathematik
der Technischen Universität Darmstadt
zur Erlangung des Grades eines
Doktors der Naturwissenschaften
(Dr. rer. nat.)
genehmigte

Dissertation

von

Dipl.-Math. (Wirtschaft) Birgit Debrabant, geb. Niese
aus Dresden

Referent:	Prof. Dr. J. Lehn
Koreferent:	Prof. Dr. J. Mecke
Koreferent:	Prof. Dr. W. Stannat
Tag der Einreichung:	24. Januar 2008
Tag der mündlichen Prüfung:	29. April 2008

Darmstadt 2008
D17

Bibliografische Information der Deutschen Nationalbibliothek

Die Deutsche Nationalbibliothek verzeichnet diese Publikation in der
Deutschen Nationalbibliografie; detaillierte bibliografische Daten sind
im Internet über http://dnb.d-nb.de abrufbar.

ISBN 978-3-8325-1959-9

Logos Verlag Berlin GmbH
Comeniushof, Gubener Str. 47,
10243 Berlin
Tel.: +49 030 42 85 10 90
Fax: +49 030 42 85 10 92
INTERNET: http://www.logos-verlag.de

Acknowledgements

First of all I express my gratitude towards my supervisor Prof. Dr. Jürgen Lehn in particular for his steady encouragement, support and advice. I am grateful to Prof. Dr. Joseph Mecke (FSU Jena), Dr. Lothar Partzsch (TU Dresden) as well as Prof. Dr. Wilhelm Stannat (TU Darmstadt) for further fruitful and inspiring discussions and comments and their selfless willingness to occupy themselves with my research. I thank Prof. Dr. Erhard Cramer (RWTH Aachen) for having introduced me to the theory of generalized order statistics which inspired me to choose the subject of this thesis. I would like to thank my colleagues in the Stochastics group of the Darmstadt University of Technology for the pleasant working atmosphere in the last years. Last but not least, I thank my husband Dr. Kristian Debrabant for his never ending encouragement, advice and trust in me.

Contents

Chapter 1

Introduction

Mixed Poisson processes are a well known class of point processes derived from (stationary) Poisson processes. In particular they cover cases where the intensity of a Poisson process is unknown but can be assumed to follow a known probability distribution. This situation is common e. g. in insurance mathematics where for instance the number of accident claims in which an individual is involved and which is evolving over some time can in principal be well described by a Poisson process with an individual, yet normally unknown intensity corresponding to the individual's accident proneness. Modelling this intensity as a random variable naturally leads to a mixed model. Usually, an insurance company will have a good estimate of the associated mixing distribution due to its large portfolio of policies.

Mixed Poisson processes have been widely studied until now and there exists a largely developed theory. For a comprehensive overview of their properties we refer to the monograph of Grandell (1997). Mixed Poisson processes show a very even behavior in a sense which is reflected by the following conditional uniformity property of their occurrence times:

Provided that a certain number of points occur in a time period $[0, t]$, these process points are independent and uniformly distributed in $[0, t]$. In other words, given that a mixed Poisson process N has attained state $n \in \mathbb{N}$ at time $t > 0$, $N_t = n$, the (ordered) process points $T_1, \ldots, T_n$ occurring before t follow the joint density

$$f_{T_1, \ldots, T_n | N_t = n}(t_1, \ldots, t_n) = n! \, t^{-n}, \qquad 0 < t_1 \leq \cdots \leq t_n \leq t. \qquad (1.1)$$

According to Nawrotzki (1962) this conditional uniformity property characterizes mixed Poisson processes. Clearly, one can describe mixed Poisson processes in terms of the ordered structure of the process points. Recent attempts in this direction have been made by Hayakawa (2000) and Shaked et al. (2004). The deduced properties reflect again the regularity of mixed Poisson processes as for instance Hayakawa proves that the distribution of normalized occurrence times $\frac{T_1}{T_n}, \ldots, \frac{T_{n-1}}{T_n}$, $n \in \mathbb{N}$, is independent of the mixing distribution and goes back to ordered independent uniform distributions.

Nevertheless, it is still of interest to find other classes of point processes in order to be able to model a more fluctuating behavior. There has already been done a lot of work and numerous generalizations of Poisson processes can be found in the literature such as renewal processes, nonstationary Poisson processes or Cox processes. In particular, attacking the structure of the mentioned conditional distributions (1.1) of the process points is an established attempt to generalize. Results in this direction relying on order statistics based on arbitrary probability distributions were achieved by Feigin (1979) and Puri (1982). However, they mainly lead to mixed Poisson processes undergoing some time transformation.

All in all, known results in the field of (generalizations of) mixed Poisson processes and ordered structures have in common that they only refer to models of ordered random variables generated by independent, identically distributed random variates. This is surmounted by the present thesis, which introduces and studies the class of generalized order statistic processes (GOS-processes) and its properties. The distribution of these point processes on $[0, \infty)$ is specified by means of the conditional distribution of the ordered process points. Based on a sequence $\{\alpha_i\}_{i \in \mathbb{N}}$ of real parameters with $\gamma_i = \sum_{j=1}^{i}(\alpha_j + 1) > 0$ for $i \in \mathbb{N}$, a generalized order statistic process is such that, provided the process attains state $n \in \mathbb{N}$ at time $t > 0$, the (ordered) process points occurring before t follow the joint density

$$f_{T_1,\ldots,T_n \mid N_t = n}(t_1, \ldots, t_n) \;\; = \;\; t^{-\gamma_n} \prod_{i=1}^{n} \gamma_i t_i^{\alpha_i}, \qquad 0 < t_1 \leq \cdots \leq t_n \leq t. \quad (1.2)$$

This feature clearly extends the conditional uniformity property of mixed Poisson processes, which corresponds to the constant parametrizing sequence $\alpha_i = 0$ for $i \in \mathbb{N}$. Now, including a sequence $\{\alpha_i\}_{i \in \mathbb{N}}$ of parameters leads to point processes with a much more variable behavior. Moreover, the chosen model can in general not be traced back to models of ordered random variables generated by independent and identically distributed random variates. This is what constitutes the main difference between the approach chosen in the present thesis and the cited articles.

The major concern of the present work is to determine and analyze the (unconditional) distribution of GOS-processes and to elaborate in which way the defining conditional structure transfers to completely specified point processes. In analogy to mixed Poisson processes, we particularly expect to discover a mixed structure. Except for some special cases, in general it is not straight away evident how to specify the unconditional distribution of GOS-processes. Thus, we will put much effort into developing techniques appropriate for solving this problem. This results in the complete description of certain GOS-processes. In particular, after some direct approaches, we formulate the problem in various contexts in order to tackle it. So, we successfully convert the question into an analytic problem concerning the existence of special families of functions. A further approach from the viewpoint of birth processes results in a characteristic recursion for the birth rates of a GOS-process.

Concerning the organization of the document:

Chapter 2 provides background knowledge. Besides a generalized model of order statistics, we introduce our notion of a point process, define (mixed) Poisson and Cox processes and present several properties necessary for later purposes.

Chapter 3 introduces the central concepts and questions of this thesis: Firstly, we present in greater detail known results which connect mixed Poisson processes with models of ordered random variables, i. e. with order statistics. Thereafter we introduce the class of generalized order statistic processes, which are only specified by the conditional distribution of the process points given by (1.2). What follows is a brief analysis of the parametrized density models involved in the definition. These are connected to generalized order statistics, which are well known in the literature, compare Kamps (1995). We further point out basic analytical properties of GOS-processes before we give a representation concerning the (unconditional) distribution of a GOS-process. So, Theorem 32 achieves a characterization in terms of the (unconditional) distribution of the process points. We discover joint densities of the form $\prod_{i=1}^{n} t_i^{\alpha_i} \varphi_n(t_n)$, where $\{\varphi_n\}_{n\in\mathbb{N}}$ is some family of functions. Based on Theorem 32 and $\{\varphi_n\}_{n\in\mathbb{N}}$, Section 3.5 presents basic distributional properties of GOS-processes. We deduce several formulas expressing fundamental probabilities in terms of $\{\varphi_n\}_{n\in\mathbb{N}}$ such as the probability that a GOS-process stays in state n at time t. Moreover, we show in analogy to Hayakawa (2000) that successive process points can be normalized such that their joint distribution is independent of $\{\varphi_n\}_{n\in\mathbb{N}}$. Strictly speaking we find

$$f_{\frac{T_1}{T_n},\dots,\frac{T_{n-1}}{T_n}}(t_1,\dots,t_{n-1}) = \prod_{i=1}^{n-1} \gamma_i t_i^{\alpha_i}, \qquad 0 < t_1 \leq \cdots \leq t_{n-1} \leq 1,\ n \in \mathbb{N},$$

which reflects the generalized order statistic structure. The concluding Section 3.6 studies whether and when GOS-processes possibly explode.

Chapter 4 is dedicated to GOS-processes with respect to constant and eventually constant parametrizing sequences. Theorem 40 of Section 4.2 gives a characterization in the case of eventually constant parametrizing sequences. This result specifies the distribution of appropriate GOS-processes, but at the same time reveals that the distribution is not uniquely determined by the parametrizing sequence. Instead and as expected, we discover a mixed structure paralleling the one of mixed Poisson processes. In addition, Theorem 40 finally enables us to study some examples beyond mixed Poisson processes. It turns out that particular delayed renewal processes satisfy the generalized order statistic property. Section 4.3 highlights the crucial fact that there are sequences, which cannot be associated to proper GOS-processes although yielding proper conditional distributions. In Section 4.4, inspired by the example of delayed renewal processes, which in our case actually are Poisson processes of which a certain number of points are deleted at the beginning, we deduce a decomposition of GOS-processes based on mixed Poisson processes. This result only holds for parametrizing sequences consisting of natural numbers as it refers to the discrete process of deleting points. As a corollary we deduce an asymptotic property. Preparing studies of the subsequent chapter, Section 4.5 presents a detailed analysis of the distributions appearing in the case of eventually constant parametrizing sequences in order to find more advantageous representations.

Chapter 5 returns to general parametrizing sequences. We introduce the concept of generators in order to specify distributions of GOS-processes with respect to sequences beyond those eventually constant:

As projective family, the distributions of the first n successive points of a GOS-process for varying n are interrelated and with them so are the functions $\{\varphi_n\}_{n \in \mathbb{N}}$. In principal, to know a single of these functions together with the parametrizing sequence suffices to completely describe the process' distribution. A generator is such a single function. This concept enables us to find GOS-processes with respect to nonnegative, mainly increasing convergent and periodic sequences. However, generators are not unique, that is, there exist different point processes satisfying the generalized order statistic property with respect to a given sequence. Thus, we dedicate Section 5.5 to the question which connections can be found between different generators associated to one sequence. With the help of martingales we achieve a characterization of all those GOS-processes associated to a single given sequence. This results in some interesting martingales linked to mixed Poisson processes.

Chapter 6 aims at the intrinsic question whether generalized order statistic processes belong to other known classes of point processes. Under this aspect, we consider birth and Cox processes. Concerning the class of birth processes, we observe that GOS-processes are contained therein and can be characterized by a recursion holding for their birth rates. What concerns Cox processes, it turns out that only in rare cases GOS-processes are Cox processes simultaneously. We completely determine which GOS-processes with respect to eventually constant parametrizing sequences are Coxian.

The concluding chapter points out several open problems related to GOS-processes.

The results presented in this thesis are the work of the author if not mentioned otherwise. As a matter of fact, in the preliminary Chapter 2 we state mostly known results. However, Proposition 11 and Lemma 12 are the author's work. Proposition 36 and its proof have been suggested by Dr. L. Partzsch from the Technical University of Dresden, Lemma 63 and its proof by Prof. Dr. W. Stannat from the Darmstadt University of Technology. In Chapter 6 the proofs of Proposition 76 and Theorem 78 as well as Lemma 77, further, topic and results of Section 4.4 have been inspired by Prof. Dr. J. Mecke from the Friedrich Schiller University of Jena.

Chapter 2

Preliminaries

This chapter provides preliminary and known results and notations referred to in the present thesis. We will introduce generalized order statistics, the notion of stochastic processes and present known point processes related to this work.

Firstly, note that the integrals used in this work are Riemann as well as Lebesgue integrals and integrals with respect to (w.r.t.) arbitrary measures. Thereby, we will often transform one notion of integrals into another without mentioning it. Several theorems justifying the integral calculus used here are presented in Appendix C.

By $\mathbb{1}_A$ we denote the indicator function corresponding to a set A. Moreover, let $\mathcal{B}^n$ denote the Borelian sets on $\mathbb{R}^n$ and ℓ^n the Lebesgue measure on $\mathcal{B}^n$, $n \in \mathbb{N} = \{1, 2, 3, \ldots\}$. For two measurable spaces $(X_1, \mathcal{F}_1)$, $(X_2, \mathcal{F}_2)$, a measure μ on $\mathcal{F}_1$ and a measurable mapping $T : X_1 \to X_2$ the measure induced by T (w.r.t. μ), $T(\mu)$, is the measure on $\mathcal{F}_2$ given by $T(\mu)(A) = \mu(T^{-1}(A))$ for $A \in \mathcal{F}_2$, where $T^{-1}(A) = \{x \in X_1 | T(x) \in A\}$. We consider further a probability space $(\Omega, \mathcal{F}, P)$. If we say an n-dimensional random vector follows a (joint) density, we mean the usual density w.r.t. ℓ^n. Basic probability densities w.r.t. the Lebesgue measure on $\mathbb{R}$ used in the sequel can be found in Appendix D. For a distribution function F we denote by F^{-1} its pseudo-inverse function that is $F^{-1}(y) = \inf\{x \in \mathbb{R} | F(x) \geq y\}$ for $y \in \mathbb{R}$.

2.1 Generalized order statistics

Generalized order statistics are a generalization of classical ordinary order statistics and were originally studied by Kamps (1995). Except for slight differences concerning the chosen parametrization he gives the subsequent definition.

For $n \in \mathbb{N}$ and $t > 0$ let us denote by K_n and $K_n(t)$ the sets

$$
\begin{aligned}
K_n &= \{(s_1, \ldots, s_n) \in (0, \infty)^n | 0 < s_1 \leq \cdots \leq s_n\}, \\
K_n(t) &= \{(s_1, \ldots, s_n) \in (0, \infty)^n | 0 < s_1 \leq \cdots \leq s_n \leq t\}.
\end{aligned}
$$

Definition 1: *Let $n \in \mathbb{N}$ and $a_1, \ldots, a_n \in \mathbb{R}$ be such that $g_i = \sum_{j=i}^{n}(a_j + 1) > 0$, for $i = 1, \ldots, n$. If the random variables $U_{i:n}$, $i = 1, \ldots, n$, follow the joint density*

$$f_{U_{1:n},\ldots,U_{n:n}}(u_1, \ldots, u_n) = \prod_{i=1}^{n} \left(g_i \left(1 - u_i\right)^{a_i} \right) \cdot \mathbb{1}_{K_n(1)}(u_1, \ldots, u_n), \quad u_1, \ldots, u_n \in \mathbb{R},$$

then they are called **uniform generalized order statistics**.

For a distribution function F the random variables $X_{i:n}$, $i = 1, \ldots, n$, defined by

$$X_{i:n} = F^{-1}(U_{i:n}), \qquad i = 1, \ldots, n,$$

are called **generalized order statistics based on F**.

If $a_i = 0$ for $i = 1, \ldots, n$ we call $U_{i:n}$ resp. $X_{i:n}$, $i = 1, \ldots, n$, **ordinary order statistics (based on F)**.

Note that if F has the density f w. r. t. ℓ, generalized order statistics $X_{1:n}, \ldots, X_{n:n}$, $n \in \mathbb{N}$, based on F follow the density

$$f_{X_{1:n},\ldots,X_{n:n}}(x_1, \ldots, x_n) = \prod_{i=1}^{n} \left[g_i f(x_i) \left(1 - F(x_i)\right)^{a_i} \right], \tag{2.1}$$

for $x_1, \ldots, x_n \in \mathbb{R}$ with $x_1 \leq \cdots \leq x_n$.

Cramer (2002) presented a constructive way to obtain generalized order statistics. In his Definition 3.1.5. and the preceding discussion we essentially find the following:

Let the assumptions of Definition 1 hold. Let further $B_1, \ldots, B_n$ be independent Beta distributed random variables such that $B_i \sim B(g_i, 1)$, i.e. B_i follows the density $f_{B_i}(x) = g_i x^{g_i - 1} \mathbb{1}_{(0,1)}(x)$, $x \in \mathbb{R}$, $i = 1, \ldots, n$. Then the random variables

$$F^{-1}\left(1 - \prod_{i=1}^{j} B_i \right), \qquad j = 1, \ldots, n, \tag{2.2}$$

are distributed like generalized order statistics $X_{1:n}, \ldots, X_{n:n}$ based on F, $n \in \mathbb{N}$.

Ordinary order statistics can be constructed by ordering n independent, identically according to F distributed random variables beginning with the smallest. If F has the density f then $n \in \mathbb{N}$ ordinary order statistics $X_{1:n}, \ldots, X_{n:n}$ follow the density

$$f_{X_{1:n},\ldots,X_{n:n}}(x_1, \ldots, x_n) = n! f(x_1) \cdots f(x_n), \qquad x_1 \leq x_2 \leq \cdots \leq x_n. \tag{2.3}$$

In analogy to the model of generalized order statistics Burkschat et al. (2003) introduced the notion of dual generalized order statistics:

Definition 2: *Let $n \in \mathbb{N}$ and $a_1, \ldots, a_n \in \mathbb{R}$ be such that $g_i = \sum_{j=i}^{n}(a_j + 1) > 0$ for $i = 1, \ldots, n$. If the random variables $U_{i:n}^d$, $i = 1, \ldots, n$, follow the joint density*

$$f_{U_{1:n}^d,\ldots,U_{n:n}^d}(u_1, \ldots, u_n) = \prod_{i=1}^{n} g_i u_i^{a_i} \cdot \mathbb{1}_{K_n(1)}(u_n, \ldots, u_1), \tag{2.4}$$

where $u_1, \ldots, u_n \in \mathbb{R}$, *then they are called* **uniform dual generalized order statistics**.

For a distribution function F the random variables $X_{i:n}^d$, $i = 1, \ldots, n$, defined by

$$X_{i:n}^d = F^{-1}(U_{i:n}^d), \qquad i = 1, \ldots, n,$$

are called **dual generalized order statistics based on** F.

Burkschat et al. (2003) further point out the following formula expressing the distribution of the i-th dual generalized order statistics $X_{i:n}^d$, $i \leq n$, based on F in terms of a Meijer's G-function, a special function which can be defined by a complex contour integral, cp. Appendix A.3:

$$P(X_{i:n}^d \leq x) = \left(\prod_{j=1}^{i} g_j\right) \int_0^{F(x)} \mathrm{G}_i[y \mid g_1, \ldots, g_i]\, dy, \qquad x \in \mathbb{R}. \tag{2.5}$$

2.2 Point processes

In the following sections the reader finds a survey of stochastic processes and point processes on $[0, \infty)$ in particular referring to mixed Poisson and Cox processes. Besides the definitions we present essential features and basic formulas.

Let $\mathbb{R}^{[0,\infty)}$ be the set of real functions on $[0, \infty)$ endowed with the σ-algebra $\mathcal{H}\left(\mathbb{R}^{[0,\infty)}\right)$ generated by

$$\left\{f \in \mathbb{R}^{[0,\infty)} \mid f(t) \in A\right\}, \qquad t \geq 0,\ A \in \mathcal{B}.$$

Definition 3: *A* **stochastic process** *X is a measurable mapping from $(\Omega, \mathcal{F}, P)$ into $\left(\mathbb{R}^{[0,\infty)}, \mathcal{H}\left(\mathbb{R}^{[0,\infty)}\right)\right)$. The distribution P_X of X is the measure on $\mathcal{H}\left(\mathbb{R}^{[0,\infty)}\right)$ induced by X.*

For $t \geq 0$ we denote by X_t or $X(t)$ the state of X at time t, that is, the random variable $X_t : \omega \mapsto X(\omega)(t)$. Further, let $\mathcal{H}_t^X = \sigma(X_s : s \leq t)$ be the natural history of X at time $t \geq 0$.

Definition 4: *A stochastic process X is a* **Markov process***, if*

$$P\left(X_t \in B \mid \mathcal{H}_s^X\right) = P(X_t \in B \mid X_s) \qquad P\text{-a. s.} \tag{2.6}$$

for all $0 \leq s \leq t$ and $B \in \mathcal{B}$.

Thereby equation (2.6) holds if and only if

$$P(X_t \in B \mid X_{s_1}, \ldots, X_{s_n}) = P(X_t \in B \mid X_{s_n}) \qquad P\text{-a. s.}$$

for all $n \in \mathbb{N}$, $0 \leq s_1 \leq \cdots \leq s_n \leq t$.

In analogy to the definition of Markov processes a sequence $\{X_i\}_{i\in\mathbb{N}}$ of random variables is called a **Markov chain** if

$$P(X_{n+1} \in B | X_1, \ldots, X_n) = P(X_{n+1} \in B | X_n), \qquad n \in \mathbb{N},\, B \in \mathcal{B}.$$

The concept of stationarity addresses the invariance of the distribution of stochastic processes with respect to time shifts:

Definition 5: *A stochastic process X has* **stationary increments** *if for every $n \in \mathbb{N}$, $0 \le s_1 < t_1 \le s_2 < t_2 \le \cdots \le s_n < t_n$ and $h \ge 0$ the joint distribution of the increments $X_{t_1+h} - X_{s_1+h}, \ldots, X_{t_n+h} - X_{s_n+h}$ does not depend on h.*

Point processes are stochastic processes with values in $\mathbb{N}_0 \cup \{\infty\}$, where $\mathbb{N}_0 = \{0, 1, 2, \ldots\}$:

Consider the measurable space $(\mathcal{N}, \mathcal{H}(\mathcal{N}))$, where $\mathcal{N}$ is the set of all simple counting functions $\mathfrak{n} : [0, \infty) \to \mathbb{N}_0 \cup \{\infty\}$, i.e.,

  (i) $\mathfrak{n}(0) = 0$,

  (ii) $\mathfrak{n}$ is increasing and right-continuous, and

  (iii) $\mathfrak{n}(t) - \lim_{h\to 0, h>0} \mathfrak{n}(t - h) \in \{0, 1\}$ for $t > 0$,

and where $\mathcal{H}(\mathcal{N})$ denotes the σ-algebra generated by the cylindric sets $\{\mathfrak{n} \in \mathcal{N} \,|\, \mathfrak{n}(t) = n\}$, $n \in \mathbb{N}_0$, $t \ge 0$.

Definition 6: *A (simple)* **point process** *N is a measurable mapping from $(\Omega, \mathcal{F}, P)$ into $(\mathcal{N}, \mathcal{H}(\mathcal{N}))$.*

There are two alternative ways to describe a point process:

  a) by the sequence of its process points/occurrence times $T_1, T_2, T_3 \ldots$, where

$$T_i = \inf\{t \ge 0 : N_t = i\}, \qquad i \in \mathbb{N}, \tag{2.7}$$

    and where $\inf \emptyset = \infty$, or

  b) by its sojourn times $S_1, S_2, \ldots$, the time periods meanwhile the process stays in one state:

$$S_i = \begin{cases} T_1 & \text{for } i = 1, \\ T_i - T_{i-1} & \text{for } i > 1 \text{ and if } T_{i-1} < \infty. \end{cases} \tag{2.8}$$

In the sequel, we will call a point process trivial, if almost surely no jumps occur, that is if $P(T_1 < \infty) = 0$.

Under the assumptions of our point process model it is possible that infinitely many jumps occur in finite time. Therefore, we consider the event

$$\left\{ \sup_{n \in \mathbb{N}} T_n < \infty \right\},$$

which is called explosion, cp. Schmidt (1996). We say that a point process N explodes with a positive probability if $P(\sup_{n \in \mathbb{N}} T_n < \infty) > 0$. Then the following lemma holds:

Lemma 7: *Let N be a point process, then the following statements are equivalent:*

(i) Explosion occurs with a positive probability, i. e.

$$P\left(\sup_{n \in \mathbb{N}} T_n < \infty \right) > 0.$$

(ii) There exists a $t > 0$ such that $\sum_{n=0}^{\infty} P(N_t = n) < 1$.

Proof: This is a direct consequence of Lemma 2.1.4, Schmidt (1996). ∎

Definition 8: *A Markovian point process N is a* **birth process** *with* **transition intensities/birth rates** κ_n *if*

$$P(N_{t+h} = m | N_t = n) = \begin{cases} 1 - \kappa_n(t)h + o(h), & m = n, \\ \kappa_n(t)h + o(h), & m = n + 1, \qquad \text{as } h \downarrow 0, \\ o(h), & m > n + 1, \end{cases}$$

where $t \geq 0$, $h > 0$, $n, m \in \mathbb{N}_0$ such that $P(N_t = n) > 0$ and where o is a function such that $\lim_{h \to 0} \frac{o(h)}{h} = 0$.

For a birth process N we find

$$P(N_{t+h} = n) = P(N_t = n)(1 - \kappa_n(t)h + o(h)) + P(N_t = n - 1)(\kappa_{n-1}(t)h + o(h))$$
$$+ \sum_{k=0}^{n-2} P(N_t = k)o(h)$$

which results in

$$\frac{d}{dt} P(N_t = 0) = -\kappa_0(t)P(N_t = 0), \tag{2.9}$$

$$\frac{d}{dt} P(N_t = n) = -\kappa_n(t)P(N_t = n) + \kappa_{n-1}(t)P(N_t = n - 1) \tag{2.10}$$

if the transition intensities are continuous, cf. Grandell (1997).

2.2.1   Stationary Poisson processes

We recall the definition of Poisson processes on $[0, \infty)$ and some of their distributional properties:

Definition 9: *A point process N is a* **(stationary) Poisson process with intensity** $\lambda \geq 0$, *say* $N \sim P^\lambda$, *if it has independent and Poisson distributed increments such that*

$$N_t - N_s \sim \text{Poi}((t - s)\lambda), \qquad 0 \leq s < t.$$

Compare Appendix D for Poisson distributions. For the finite-dimensional distributions of a Poisson process at $n \in \mathbb{N}$ times $t_1, \ldots, t_n$, where $0 = t_0 \leq t_1 < \ldots < t_n$, we obtain

$$P(N_{t_1} = k_1, \ldots, N_{t_n} = k_n)$$

$$= \ P(N_{t_1} - N_0 = k_1, N_{t_2} - N_{t_1} = k_2 - k_1, \ldots, N_{t_n} - N_{t_{n-1}} = k_n - k_{n-1})$$

$$= \ P(N_{t_1} - N_0 = k_1) \cdot P(N_{t_2} - N_{t_1} = k_2 - k_1) \cdots P(N_{t_n} - N_{t_{n-1}} = k_n - k_{n-1})$$

$$= \begin{cases} \lambda^{k_n} e^{-\lambda t_n} \displaystyle\prod_{l=1}^{n} \frac{(t_l - t_{l-1})^{k_l - k_{l-1}}}{(k_l - k_{l-1})!} & \text{for } 0 = k_0 \leq k_1 \leq \ldots \leq k_n, \\ 0 & \text{else,} \end{cases} \qquad (2.11)$$

where $k_1, \ldots, k_n \in \mathbb{N}_0$.

It is well known that the sojourn times $S_1, S_2, \ldots$ of a Poisson process with positive intensity $\lambda > 0$ are independent and identically $\text{Exp}(\lambda)$-distributed, compare for instance (Schmidt, 1996, Theorem 2.3.4). This yields further that its occurrence times follow the densities

$$f_{T_n}(t) \ = \ \frac{t^{n-1}}{(n-1)!} \lambda^n e^{-\lambda t}, \qquad\qquad t > 0, \qquad (2.12)$$

$$f_{T_1, \ldots, T_n}(t_1, \ldots, t_n) \ = \ \lambda^n e^{-\lambda t_n}, \qquad n \in \mathbb{N}, \, 0 < t_1 \leq \cdots \leq t_n. \qquad (2.13)$$

Note that a Poisson process with intensity $\lambda = 0$ remains in its initial state 0 with probability 1.

2.2.2   Mixed Poisson processes

Definition 10: *A point process N is a* **mixed Poisson process**, *if its distribution P_N verifies*

$$P_N(A) \ = \ \int_{[0,\infty)} P^\lambda(A) \, dV(\lambda), \qquad A \in \mathcal{H}(\mathcal{N}),$$

where P^λ *denotes the distribution of a (stationary) Poisson process with intensity λ and where V is a probability distribution on $[0, \infty)$ called* **mixing distribution**.

This class of processes has been studied in detail basically for the first time by Lundberg (1940) resp. Lundberg (1964); a detailed survey of the until now developed theory can be found in Grandell (1997).

For the joint distribution of $n \in \mathbb{N}$ states at times $t_1, \ldots, t_n \in \mathbb{R}$ such that $0 = t_0 < t_1 < \cdots < t_n$ we find

$$P(N_{t_1} = k_1, \ldots, N_{t_n} = k_n) \overset{(2.11)}{=} \prod_{l=1}^{n} \frac{(t_l - t_{l-1})^{k_l - k_{l-1}}}{(k_l - k_{l-1})!} \int_{[0,\infty)} \lambda^{k_n} e^{-\lambda t_n} \, dV(\lambda)$$

$$= \frac{k_n!}{t_n^{k_n}} \prod_{l=1}^{n} \frac{(t_l - t_{l-1})^{k_l - k_{l-1}}}{(k_l - k_{l-1})!} \int_{[0,\infty)} \frac{(\lambda t_n)^{k_n}}{k_n!} e^{-\lambda t_n} \, dV(\lambda)$$

$$= \frac{k_n!}{t_n^{k_n}} \prod_{l=1}^{n} \frac{(t_l - t_{l-1})^{k_l - k_{l-1}}}{(k_l - k_{l-1})!} P(N_{t_n} = k_n), \tag{2.14}$$

where $k_1, \ldots, k_n \in \mathbb{N}_0$ such that $0 \leq k_1 \leq \cdots \leq k_n$. Especially for $n = 1$ holds

$$P(N_t = k) = \int_{[0,\infty)} \frac{(\lambda t)^k}{k!} e^{-\lambda t} \, dV(\lambda) = \frac{(-t)^k}{k!} \hat{V}^{(k)}(t), \qquad t > 0, \, k \in \mathbb{N}_0, \tag{2.15}$$

where $\hat{V}^{(k)}$ denotes the k-th derivative of the Laplace transform $\hat{V}$ of V. Further, we obtain

$$P(N_{t_1} = k_1, \ldots, N_{t_{n-1}} = k_{n-1} | N_{t_n} = k_n) = \frac{P(N_{t_1} = k_1, \ldots, N_{t_n} = k_n)}{P(N_{t_n} = k_n)}$$

$$= \frac{k_n!}{t_n^{k_n}} \prod_{l=1}^{n} \frac{(t_l - t_{l-1})^{k_l - k_{l-1}}}{(k_l - k_{l-1})!}, \tag{2.16}$$

which is independent of the mixing distribution V, and

$$P(N_{t_n} = k_n | N_{t_1} = k_1, \ldots, N_{t_{n-1}} = k_{n-1}) = \frac{P(N_{t_1} = k_1, \ldots, N_{t_n} = k_n)}{P(N_{t_1} = k_1, \ldots, N_{t_{n-1}} = k_{n-1})}$$

$$= \frac{(t_n - t_{n-1})^{k_n - k_{n-1}}}{(k_n - k_{n-1})!} \cdot \frac{\int_{[0,\infty)} \lambda^{k_n} e^{-\lambda t_n} dV(\lambda)}{\int_{[0,\infty)} \lambda^{k_{n-1}} e^{-\lambda t_{n-1}} dV(\lambda)} \tag{2.17}$$

for $n \in \mathbb{N}$, $k_1, \ldots, k_n \in \mathbb{N}_0$ with $0 = k_0 \leq k_1 \leq \cdots \leq k_n$ and $t_1, \ldots, t_n \in \mathbb{R}$ such that $0 = t_0 \leq t_1 < \ldots < t_n$. Hence, mixed Poisson processes are Markovian. For $n = 2$, expression (2.17), generally, does not depend on the time difference $t_2 - t_1$ only, but also on t_1, t_2 which implies that mixed Poisson processes in general are nonhomogeneous, i.e.

$$P(N_{t_2} = k_2 | N_{t_1} = k_1) \neq P(N_{t_2+h} = k_2 | N_{t_1+h} = k_1)$$

is valid for some $h > 0$.

Concerning the asymptotic behavior of a mixed Poisson process, we find that the ratio $\frac{n}{T_n}$ converges in distribution to a random variable following the mixing distribution. Defining a mixed Poisson process alternatively by $t \mapsto N(\Lambda \cdot t)$, where $N \sim \mathrm{P}^1$ is a standard Poisson process and Λ a nonnegative random variable, cp. Grandell (1997), we had moreover $\frac{n}{T_n} \to \Lambda$ absolute surely for $n \to \infty$ due to the additional structure.

Whereas the occurrence times of a stationary Poisson process with positive intensity are absolutely continuous w.r.t. the Lebesgue measure, in the case of mixed Poisson processes the process might remain in its initial state 0 with a positive probability, namely if the mixing distribution assigns a positive weight to 0. Then the occurrence times are infinite. However, given that at least one jump occurs, the process jumps infinitely often almost surely and the occurrence times are absolutely continuous w.r.t. the Lebesgue measure. Let us introduce some notations before we specify and summarize the just said in Proposition 11.

By δ_x for some number x we mean Dirac's measure, i.e. $\delta_x(A) = 1$ if $x \in A$ or 0 else for a set A. For some measure μ we denote by $f \odot \mu$ the measure that is absolutely continuous with respect to μ and with Radon-Nikodým-density f.

Moreover, let $\bar{\mathbb{R}}$ denote the extended real line that is $\bar{\mathbb{R}} = \mathbb{R} \cup \{-\infty, \infty\}$ and put especially $(\pm\infty) \cdot 0 = 0 \cdot (\pm\infty) = 0$. Further let $\bar{\mathcal{B}}$ extend the Borelian σ-algebra $\mathcal{B}$ such that

$$\bar{\mathcal{B}} = \left\{ B \cup E \mid B \in \mathcal{B}, E \subset \{-\infty, \infty\} \right\},$$

and let $\bar{\mathcal{B}}^n = \bar{\mathcal{B}} \otimes \cdots \otimes \bar{\mathcal{B}}$ be the corresponding product-σ-algebra in $\bar{\mathbb{R}}^n$. Let $\bar{\ell}^n$ be the measure that extends ℓ^n to $\bar{\mathcal{B}}^n$ such that $\bar{\ell}^n(\cdot) = \ell^n(\cdot \cap \mathbb{R}^n)$. By a (probability) density on $\bar{\mathbb{R}}^n$ (w.r.t. $\bar{\ell}^n$), say $\bar{f}$, we mean a nonnegative $(\bar{\mathcal{B}}^n, \bar{\mathcal{B}})$-measurable function $\bar{f} : \bar{\mathbb{R}}^n \to [0, \infty]$ such that $\bar{f} \odot \bar{\ell}^n$ is a (probability) measure on $(\bar{\mathbb{R}}^n, \bar{\mathcal{B}}^n)$. Since $\bar{\ell}^n(\bar{\mathbb{R}}^n \backslash \mathbb{R}^n) = \ell^n\left((\bar{\mathbb{R}}^n \backslash \mathbb{R}^n) \cap \mathbb{R}^n\right) = 0$, the measure $\bar{f} \odot \bar{\ell}^n$ is concentrated on $\mathbb{R}^n$ and the values of $\bar{f}$ on $\bar{\mathbb{R}}^n \backslash \mathbb{R}^n$ do not contribute to the measure of a set. Moreover, the restriction of $\bar{f}$ on $\mathbb{R}^n$ is the density of a (probability) measure on $(\mathbb{R}^n, \mathcal{B}^n)$. On the other hand, each density f on $\mathbb{R}^n$ (w.r.t. ℓ^n) can be extended to a density $\bar{f}$ on $\bar{\mathbb{R}}^n$ (w.r.t. $\bar{\ell}^n$) such that $\bar{f} \odot \bar{\ell}^n = f \odot \ell^n$ on $\mathcal{B}^n$. The values of $\bar{f}$ on $\bar{\mathbb{R}}^n \backslash \mathbb{R}^n$ do not contribute to the measure and can be chosen arbitrarily. Due to these comments, in the sequel we will implicitly identify a ℓ^n-density f on $\mathbb{R}^n$ with a $\bar{\ell}^n$-density $\bar{f}$ on $\bar{\mathbb{R}}^n$ and vice versa, as the case may be, without specifying the values of $\bar{f}$ on $\bar{\mathbb{R}}^n \backslash \mathbb{R}^n$. To omit complicated notations, we do not distinguish between f and $\bar{f}$ and especially we use f simultaneously as a density on $\mathbb{R}^n$ and as density on $\bar{\mathbb{R}}^n$ in the described sense.

For a point process N the random vector $(T_1, \ldots, T_n)$ of its first $n \in \mathbb{N}$ occurrence times is a measurable mapping from $(\Omega, \mathcal{F})$ into $(\bar{\mathbb{R}}^n, \bar{\mathcal{B}}^n)$. Thereby, $T_n = \infty$ means that there occur at most $n - 1$ jumps. We will denote by $P_{T_1, \ldots, T_n}$ the distribution of $(T_1, \ldots, T_n)$, i.e. $P_{T_1, \ldots, T_n}$ is the measure on $(\bar{\mathbb{R}}^n, \bar{\mathcal{B}}^n)$ induced by the mapping $(T_1, \ldots, T_n)$.

Proposition 11: *Let N be a mixed Poisson process with mixing distribution V. Then, the distribution $P_{T_1,\ldots,T_n}$ of $n \in \mathbb{N}$ successive occurrence times $T_1, \ldots, T_n$ of N satisfies*

$$P_{T_1,\ldots,T_n} = P(T_1 < \infty)\, f_{T_1,\ldots,T_n} \odot \bar{\ell}^n + P(T_1 = \infty)\, \delta_{(\infty,\ldots,\infty)}$$

$$= V_{>0} \cdot f_{T_1,\ldots,T_n} \odot \bar{\ell}^n + V_0 \cdot \delta_{(\infty,\ldots,\infty)},$$

where $V_{>0} = V((0,\infty)) = P(T_1 < \infty)$, $V_0 = V(\{0\}) = P(T_1 = \infty)$ and

$$f_{T_1,\ldots,T_n}(t_1,\ldots,t_n) = \int_{(0,\infty)} \lambda^n e^{-\lambda t_n}\, dW(\lambda), \qquad 0 < t_1 \leq \cdots \leq t_n,$$

with a probability distribution W absolutely continuous w. r. t. V such that

$$\frac{dW}{dV}(\lambda) = \begin{cases} \frac{1}{V_{>0}} & \text{for } \lambda > 0, \\ 0 & \text{for } \lambda = 0, \end{cases} \qquad V\text{-almost surely.}$$

Proof: As $\{T_1 = \infty\} = \cap_{t=0}^{\infty}\{N_t = 0\}$ and $\{N_t = 0\} \supset \{N_{t+1} = 0\}$ for $t \geq 0$ we find

$$P(T_1 = \infty) = \lim_{t\to\infty} P(N_t = 0) = \lim_{t\to\infty} \int_{[0,\infty)} e^{-\lambda t} dV(\lambda)$$

$$= \lim_{t\to\infty} \left(V(\{0\}) + \int_{(0,\infty)} e^{-\lambda t} dV(\lambda) \right) = V_0.$$

Further with $V_{>0} = P(T_1 < \infty)$ we obtain

$$P(T_1 \leq t_1,\ldots,T_n \leq t_n | T_1 < \infty) = \frac{\int_{[0,\infty)} P^\lambda(T_1 \leq t_1,\ldots,T_n \leq t_n) dV(\lambda)}{V_{>0}}$$

$$= \frac{V_0 \cdot P^0(T_1 \leq t_1,\ldots,T_n \leq t_n) + \int_{(0,\infty)} P^\lambda(T_1 \leq t_1,\ldots,T_n \leq t_n) dV(\lambda)}{V_{>0}}$$

$$= \frac{\int\cdots\int_B \int_{(0,\infty)} \lambda^n e^{-\lambda s_n} dV(\lambda) d(s_1,\ldots,s_n)}{V_{>0}}$$

where $n \in \mathbb{N}$, $0 < t_1 \leq \cdots \leq t_n$ and

$$B = \left\{ (s_1,\ldots,s_n) \in \mathbb{R}^n \;\middle|\; \begin{array}{ll} 0 < s_i \leq s_{i+1}, & i = 1,\ldots,n-1 \\ s_i \leq t_i, & i = 1,\ldots,n \end{array} \right\}.$$

That is why, provided $T_1 < \infty$, the occurrence times follow the probability density

$$f_{T_1,\ldots,T_n}(t_1,\ldots,t_n) = \int_{(0,\infty)} \lambda^n e^{-\lambda t_n} dW(\lambda) \cdot \mathbb{1}_{K_n}(t_1,\ldots,t_n),$$

for $n \in \mathbb{N}$ and $t_1, \ldots, t_n > 0$ and where W is a probability distribution absolutely continuous with respect to V such that

$$\frac{dW}{dV}(\lambda) = \begin{cases} \frac{1}{V_{>0}} & \text{if } \lambda > 0, \\ 0 & \text{else,} \end{cases} \qquad V\text{-a. s.}$$

Altogether we have

$$P_{T_1, \ldots, T_n}(A)$$

$$= V_{>0} \cdot P((T_1, \ldots, T_n) \in A | T_1 < \infty) + V_0 \cdot P((T_1, \ldots, T_n) \in A | T_1 = \infty)$$

$$= V_{>0} \cdot f_{T_1, \ldots, T_n} \odot \bar{\ell}^n(A) + V_0 \cdot \delta_{(\infty, \ldots, \infty)}(A), \qquad A \in \bar{\mathcal{B}}^n. \qquad \blacksquare$$

Another property which will be necessary in the sequel concerns the distribution of mixed Poisson processes with certain missing points:

Lemma 12: *Let N be a mixed Poisson process with mixing distribution V, $\gamma_0 = 0$ and $\gamma_1, \gamma_2, \ldots \in \mathbb{N}$ an increasing sequence. Let further $N^{[k]}$ for $k \in \mathbb{N}_0$ be the point process obtained after deleting for $i = 1, \ldots, k$ the process points $T_{\gamma_{i-1}+1}, \ldots, T_{\gamma_i - 1}$ if $\gamma_{i-1} + 1 \leq \gamma_i - 1$.*

Then, with W, V_0 and $V_{>0}$ as in Proposition 11, the distribution $P_{T_1^{[k]}, \ldots, T_n^{[k]}}$ of $n \geq k$ successive occurrence times of $N^{[k]}$ satisfies

$$P_{T_1^{[k]}, \ldots, T_n^{[k]}} = V_{>0} \cdot f_n^{[k]} \odot \bar{\ell}^n + V_0 \cdot \delta_{(\infty, \ldots, \infty)}, \qquad (2.18)$$

where

$$f_n^{[k]}(t_1, \ldots, t_n) = \frac{\prod_{i=1}^{k}(t_i - t_{i-1})^{\alpha_i}}{\prod_{i=1}^{k} \alpha_i!} \int_{(0, \infty)} \lambda^{\gamma_k + n - k} e^{-\lambda t_n} \, dW(\lambda) \cdot \mathbb{1}_{K_n}(t_1, \ldots, t_n) \qquad (2.19)$$

for $t_0 = 0$, $t_1, \ldots, t_n \in \mathbb{R}$, $\gamma_0 = 0$ and where $\alpha_i = \gamma_i - \gamma_{i-1} - 1$ is the number of points of N deleted between $T_{\gamma_{i-1}}$ and T_{γ_i}, $i = 2, \ldots, k$, resp. 0 and T_{γ_1} for $i = 1$.

Note that the above construction is such that for $k \in \mathbb{N}_0$ the occurrence times $T_i^{[k]}$ resp. T_i, $i \in \mathbb{N}$, of $N^{[k]}$ resp. N satisfy $T_i^{[k]} = T_{\gamma_i}$ if $i \leq k$ or $T_i^{[k]} = T_{\gamma_k + i - k}$ else.

Proof: We conduct the proof by induction on $k \in \mathbb{N}_0$:

For $k = 0$ the statement corresponds to Proposition 11. Now, let (2.18) and (2.19) be true for $k \in \mathbb{N}_0$. To obtain the point process $N^{[k+1]}$ from $N^{[k]}$ we have to delete additional α_{k+1} points of $N^{[k]}$ which are $T_{k+1}^{[k]}, \ldots, T_{k+\alpha_{k+1}}^{[k]}$ and find

$$P_{T_1^{[k+1]}, \ldots, T_n^{[k+1]}}(A) = P_{T_1^{[k]}, \ldots, T_k^{[k]}, T_{k+1}^{[k]}, \ldots, T_{n+\alpha_{k+1}}^{[k]}}(\tilde{A}),$$

where

$$\tilde{A} = B_1 \times \cdots \times B_k \times \bar{\mathbb{R}}^{\alpha_{k+1}} \times B_{k+1} \times \cdots \times B_n$$

and $A = B_1 \times \cdots \times B_n$ for some $B_1, \ldots, B_n \in \bar{\mathcal{B}}$. Note that sets A of this form generate $\bar{\mathcal{B}}^n$. This yields

$$P_{T_1^{[k]}, \ldots, T_{n+\alpha_{k+1}}^{[k]}}(\tilde{A})$$

$$\overset{(2.18)}{=} V_{>0} \cdot \int_{\tilde{A}} f_{n+\alpha_{k+1}}^{[k]} d\bar{\ell}^{n+\alpha_{k+1}} + V_0 \cdot \delta_{(\infty, \ldots, \infty)}(\tilde{A})$$

$$= V_{>0} \iint_{A\ C} f_{n+\alpha_{k+1}}^{[k]}(t_1, .., t_k, s_1, .., s_{\alpha_{k+1}}, t_{k+1}, .., t_n) d\bar{\ell}^{\alpha_{k+1}}(s_1, .., s_{\alpha_{k+1}}) d\bar{\ell}^n(t_1, .., t_n)$$

$$+ V_0 \cdot \delta_{(\infty, \ldots, \infty)}(A),$$

where $C = \bar{\mathbb{R}}^{\alpha_{k+1}}$. Since

$$\int_C f_{n+\alpha_{k+1}}^{[k]}(t_1, \ldots, t_k, s_1, \ldots, s_{\alpha_{k+1}}, t_{k+1}, \ldots, t_n) d\bar{\ell}^{\alpha_{k+1}}(s_1, \ldots, s_{\alpha_{k+1}})$$

$$\overset{(2.19)}{=} \int_{t_k}^{t_{k+1}} \int_{s_1}^{t_{k+1}} \cdots \int_{s_{\alpha_{k+1}-1}}^{t_{k+1}} \frac{\prod_{i=1}^k (t_i - t_{i-1})^{\alpha_i}}{\prod_{i=1}^k \alpha_i!} \int_{(0,\infty)} \lambda^{\gamma_k + n + \alpha_{k+1} - k} e^{-\lambda t_n} \, dW(\lambda) ds_{\alpha_{k+1}} \cdots ds_1$$

$$= \frac{\prod_{i=1}^{k+1} (t_i - t_{i-1})^{\alpha_i}}{\prod_{i=1}^{k+1} \alpha_i!} \int_{(0,\infty)} \lambda^{\gamma_{k+1} + n - (k+1)} e^{-\lambda t_n} W(\lambda), \qquad 0 < t_1 \leq \cdots \leq t_n,$$

the proof is complete.   ■

2.2.3   Nonstationary Poisson processes and Cox processes

Further generalizations of stationary Poisson processes are nonstationary Poisson processes and based thereupon Cox processes:

Definition 13: *Let $\Lambda : [0, \infty) \to [0, \infty)$ be a nondecreasing, continuous function such that $\Lambda(0) = 0$. A point process N is called **(nonstationary) Poisson process** with **intensity function** Λ, say $N \sim \mathrm{P}^\Lambda$, if its increments are independent and such that*

$$N_t - N_s \sim \mathrm{Poi}\big(\Lambda(t) - \Lambda(s)\big), \qquad 0 \leq s \leq t.$$

Remark 14:   (i) The above definition could be extended to nondecreasing, right-continuous intensity functions. Constant pieces of the intensity function result in the absence of process points at these times. Discontinuities would result in point processes with multiple points at a time that is with realizations lying outside the here considered space $\mathcal{N}$ of simple counting functions (condition (iii) would be violated). However, since the studies in the present thesis concern simple point processes only, we stick to the given definitions.

(ii) Remark further that, if $N \sim \mathrm{P}^\Lambda$ undergoes a nondecreasing, right-continuous time transformation c with $c(0) = 0$, the transformed process M, that is $M_t = N_{c(t)}$ for $t \geq 0$, satisfies $M \sim \mathrm{P}^{\Lambda \circ c}$, where $\Lambda \circ c\,(t) = \Lambda(c(t))$ for $t \geq 0$.

(iii) If Λ is boundless, the distribution P_{T_k} of the k-th occurrence time T_k, $k \in \mathbb{N}$, of a nonstationary Poisson process with intensity function Λ is absolutely continuous with respect to the Lebesgue-Stieltjes measure associated to Λ. We have

$$\frac{dP_{T_k}}{d\Lambda}(s) \;=\; \frac{\Lambda(s)^{k-1}}{(k-1)!}e^{-\Lambda(s)}, \qquad s \geq 0, \tag{2.20}$$

where Λ is to interpret as a measure on the left hand side of the above identity and as a function on the right, since

$$\int_{(a,b]} \frac{\Lambda(s)^{k-1}}{(k-1)!}e^{-\Lambda(s)}d\Lambda(s) \;\overset{Prop.C.4}{=}\; \int_{\Lambda(a)}^{\Lambda(b)} \frac{s^{k-1}}{(k-1)!}e^{-s}ds \;=\; -\sum_{i=0}^{k-1}\frac{s^i}{i!}e^{-s}\;\Big|_{\Lambda(a)}^{\Lambda(b)}$$

$$= \; P(N_a \leq k-1) - P(N_b \leq k-1) \;=\; P\big(T_k \in (a,b]\big)$$

holds for $a, b \in \mathbb{R}$ with $0 \leq a < b$. If $\Lambda(t)$ is bounded and tends to $\Lambda_\infty < \infty$ for $t \to \infty$, then P_{T_k} has an additional singular component that assigns a positive weight to $\{\infty\}$, i. e. $P(T_k = \infty) = \sum_{i=0}^{k-1}\frac{\Lambda_\infty^i}{i!}e^{-\Lambda_\infty}$.

Consider now the set $\mathcal{L}$ of nondecreasing, continuous functions $\Lambda : [0,\infty) \to [0,\infty)$ with $\Lambda(0) = 0$. Let further $\mathcal{L}$ be endowed with the σ-algebra $\mathcal{H}(\mathcal{L})$ generated by $\{\Lambda \in \mathcal{L} | \Lambda(x) \leq y\}$, $x, y \in \mathbb{R}$. In analogy to the definition of mixed Poisson processes we define Cox processes as follows, compare Grandell (1976):

Definition 15: *A point process N is a **Cox process** if its distribution P_N satisfies*

$$P_N(A) \;=\; \int_{\Lambda \in \mathcal{L}} \mathrm{P}^\Lambda(A)\,dV(\Lambda), \qquad A \in \mathcal{H}(\mathcal{N}),$$

where P^Λ is the distribution of a (nonstationary) Poisson process with intensity function Λ and V a probability distribution on $(\mathcal{L}, \mathcal{H}(\mathcal{L}))$.

Note that the distribution of a thus defined Cox process (and of mixed Poisson processes likewise) as mixture is correctly defined since the mapping $\Lambda \mapsto \mathrm{P}^\Lambda(A)$ is $\mathcal{H}(\mathcal{L})$-$\mathcal{B}$-measurable for $A \in \mathcal{H}(\mathcal{N})$, compare Lemma 1 in Grandell (1976), and due to monotone convergence P_N is again a probability measure. In Grandell (1976) we also find the following properties:

Proposition 16: (Uniqueness of the mixing distribution) *Let P_N and P_M be the distributions of two Cox processes with mixing distributions V_N and V_M, respectively. Then $P_N = P_M$ holds if and only if $V_N = V_M$.*

Proposition 17: (Time transformations) *Let N be a Cox process with mixing distribution V and let M be a time transformed process such that $M_t = N_{c(t)}$ for $t \geq 0$, where $c : [0,\infty) \to [0,\infty)$ is nondecreasing, right-continuous and such that $c(0) = 0$. Then M is a Cox process and its mixing distribution is the measure $C(V)$ on $\mathcal{H}(\mathcal{L})$ induced w. r. t. V by the mapping $C : \mathcal{L} \to \mathcal{L}$ such that $C(\Lambda) = \Lambda \circ c$.*

2.2.4 Mixed sample processes

Mixed sample processes are point processes which can be constructed as follows, compare Puri (1982):

Suppose a process is initiated with a random number Z of particles at time $t = 0$. The lifetimes $X_1, X_2, \ldots, X_Z$ of the particles are assumed to be independent and identically distributed. Then, the point process N denoting the number of particle-deaths during $(0, t]$ is called a mixed sample process. We have $N_0 = 0$ and

$$N_t = \begin{cases} 0, & \text{for } Z = 0, \\ \#\{j \in \{1, \ldots, n\} : X_j \leq t\}, & \text{for } Z = n \in \mathbb{N}, \end{cases}$$

for $t > 0$ and where $\#$ denotes the cardinality of a set.

Chapter 3

Order statistic properties and generalized order statistic processes

3.1 Mixed Poisson processes and order statistics

In this section we present several known connections between order statistics and mixed Poisson processes wherefrom the subject of the present thesis originates.

Since mixed Poisson processes were introduced as a generalization of stationary Poisson processes they have been studied intensively. In particular, there exist several characterizing properties in the literature. Well known is the following characterization of mixed Poisson processes via the conditional uniformity of their points by Nawrotzki (1962) which can be expressed as follows in terms of order statistics, compare Section 2.1:

Theorem 18: (Nawrotzki, 1962) *A point process[1] N is a mixed Poisson process if and only if, given $N_t = n$, the successive occurrence times $T_1, \ldots, T_n$ are distributed like n ordinary order statistics $X_{1:n}, \ldots, X_{n:n}$ based on a uniform distribution on $[0, t]$, for $t > 0$ and $n \in \mathbb{N}$ with $P(N_t = n) > 0$.*

The above result was part of a more comprehensive characterization of mixed Poisson processes within the class of point processes with stationary increments. However, for the mentioned characterizing property, to assume stationary increments is unnecessary, which is also affirmed by (Schmidt and Zocher, 2003, Thm. 4.2) who prove the same result in the context of claim number processes, referring to the so called multinomial property.

Moreover, Nawrotzki's original formulation was that, given n points in the interval $[0, t]$, the distribution of these points would be that of n independent random variables, each of them uniformly distributed in $[0, t]$. In this context, the n points are not to be understood as ordered points. That is why the above property is also

---

[1]Note that all theorems of Section 3.1, i.e. Theorems 18-24, assume a point process N to be such that $N_t \in \mathbb{N}_0$ for all $t \geq 0$, i.e. the case of explosion is excluded.

referred to as conditional uniformity property.

Several years later, a generalization of Nawrotzki's theorem was published by Feigin (1979) who considered arbitrary distributions at the place of uniform distributions:

Definition 19: *(i) Let $\{F_t\}_{t>0}$ be a family of distribution functions, such that the probability distribution corresponding to F_t is concentrated on $[0,t]$ for $t > 0$. We say a point process N verifies the **order statistic property** **(OS-property)** w. r. t. $\{F_t\}_{t>0}$ if, whenever $P(N_t = n) > 0$, given $N_t = n$, the successive occurrence times $T_1, \ldots, T_n$ are distributed like order statistics $X_{1:n}, \ldots, X_{n:n}$ based on F_t, $t > 0$, $n \in \mathbb{N}$.*

*(ii) The process N verifies the **uniform order statistic property (UOS-property)** if it satisfies the OS-property w. r. t. the family $\{U[0,t]\}_{t>0}$ of uniform distributions.*

Denote by $m(t) = EN_t$ for $t \geq 0$ the mean value function of N. Then the following theorem holds:

Theorem 20: (Feigin, 1979) *Let N be a point process satisfying the order statistic property w. r. t. some family of distribution functions and such that $m(t)$ is finite for $t \geq 0$ and $\lim_{t \to \infty} m(t) = \infty$. Then there exists a mixed Poisson process M such that*

$$N_t = M(m(t)), \qquad t \geq 0, \qquad a. s.$$

In the original article of Feigin (1979), the condition $\lim_{t \to \infty} m(t) = \infty$ was not mentioned though implicitly required. Its usage was afterwards pointed out by Puri (1982).

A result of Crump (1975) clarifies the relation between the distribution functions $\{F_t\}_{t>0}$ and the mean value function of the process:

Theorem 21: (Crump, 1975) *If N is a point process satisfying the OS-property w. r. t. $\{F_t\}_{t>0}$ and if $m(t)$ verifies the assumptions of Theorem 20, then*

$$F_t(x) = \frac{m(x)}{m(t)}, \qquad 0 \leq x \leq t, t > 0.$$

A relaxation of the moment conditions $m(t) < \infty$ for $t \geq 0$ and $\lim_{t \to \infty} m(t) = \infty$ has been achieved by Puri (1982). Instead, he assumes that there exists a $t_0 > 0$ such that

$$P(N_{t_0} = 0) < 1 \quad \text{and} \quad F_{t_0}(x) > 0, \quad \forall 0 < x \leq t_0, \tag{3.1}$$

and, under these milder conditions, proves the subsequent theorem:

Theorem 22: (Puri, 1982) *If N is a point process satisfying the OS-property w. r. t. $\{F_t\}_{t>0}$ and such that (3.1) holds for some $t_0 > 0$, then we have*

$$F_t(x) = \frac{q(x)}{q(t)}, \qquad 0 \leq x \leq t, t > 0,$$

where

$$q(t) = \begin{cases} F_{t_0}(t) & \text{for } t \leq t_0, \\ \frac{1}{F_t(t_0)} & \text{for } t \geq t_0. \end{cases}$$

Further, if $\lim_{t \to \infty} q(t) = \infty$, *then* N *is a mixed Poisson process* M *up to a time transformation,*

$$N_t = M(q(t)), \qquad t \geq 0, \qquad a.s.$$

Prem Puri's results go even further including the case where $\lim_{t \to \infty} q(t) < \infty$. This leads to mixed sample processes, compare Section 2.2.4.

Recent articles in the field of point processes and properties connected to order statistics were published by Hayakawa (2000) and Shaked et al. (2004). Yu Hayakawa gives an alternative proof of Theorem 18. Her approach is the ℓ_1-isotropy of the sojourn times $S_i = T_i - T_{i-1}$, $i \in \mathbb{N}$, where $T_0 = 0$, resulting from the UOS-property. Thereby, the sequence $\{S_i\}_{i \in \mathbb{N}}$ is called ℓ_1-isotropic if and only if for every $n \in \mathbb{N}$ the random variables

$$\frac{S_1}{\sum_{i=1}^n S_i}, \ldots, \frac{S_n}{\sum_{i=1}^n S_i}$$

are distributed like n independent U[0, 1]-distributed random variables $U_1, \ldots, U_n$ given $\sum_{i=1}^n U_i = 1$. Yu Hayakawa applies a result of Berman (1980) to therefrom deduce the mixed structure. Moreover, she gives another characterization of (time transformed) mixed Poisson processes based on order statistics:

Theorem 23: (Hayakawa, 2000) *For any positive integer* $n \geq 2$, *given* $T_1 < \infty$, *the normalized occurrence times*

$$\frac{T_1}{T_n}, \ldots, \frac{T_{n-1}}{T_n}$$

are distributed as the order statistics of $n-1$ *independent random variables uniformly distributed on* [0, 1] *if and only if* N *is a mixed Poisson process.*

Note that Y. Hayakawa's original Theorem 3.1. ignores the condition that at least one jump has to occur for the distribution of $\frac{T_1}{T_n}, \ldots, \frac{T_{n-1}}{T_n}$ corresponding to a mixed Poisson process to be that one specified above.

Shaked et al. (2004) present an alternative characterization of processes with the OS-property:

Theorem 24: (Shaked, Spizzichino, Suter, 2004) *Let* $c : [0, \infty) \to [0, \infty)$ *be a differentiable, strictly increasing function such that* $c(0) = 0$ *and let* $\dot{c}$ *be its first derivative. For* $t > 0$ *let* F_t *be defined by* $F_t(s) = \frac{c(s)}{c(t)}$, $0 \leq s \leq t$.

Then, a point process N *satisfies the order statistic property w. r. t.* $\{F_t\}_{t>0}$ *if and only if for every* $n \geq 1$ *such that, given* $T_n < \infty$[1]*, its first* n *occurrence times*

---

[1]Note that in Theorem 3.2. of Shaked et al. (2004), which is the one referred to, (and alike in their Theorem 2.5.) we originally find "$P(T_n < \infty) = 1$" instead of "given $T_n < \infty$" as in the above formulation.

$T_1, \ldots, T_n$ *follow a joint density of the form*

$$f_{T_1,\ldots,T_n}(t_1, \ldots, t_n) \;=\; \prod_{i=1}^{n} \dot{c}(t_i) \cdot \psi_n(c(t_n)) \mathbb{1}_{K_n}(t_1, \ldots, t_n),$$

where $t_1, \ldots, t_n \in \mathbb{R}$ *and for some function* $\psi_n : [0, \infty) \to [0, \infty)$.

So far, the cited theorems all in common apply ordinary order statistics which are based upon independently, identically distributed random variables. Now there exist various other models of ordered random variables, for instance record values of a sequence of independent and identically distributed random variables, sequential order statistics, or – as a more general model including the former two – the model of generalized order statistics. This gives rise to the question whether or not it is possible to extend the mentioned results connecting order statistics and point processes to a more general model of ordered random variables, as it is the attempt of the next section.

3.2   A generalized order statistic property

Inspired by the model of generalized order statistics, compare Section 2.1, we generalize the uniform order statistic property. In Definition 25 we introduce a class of processes satisfying a generalized order statistic property which turns out to be a rich class of point processes and which is the main object of interest in this thesis and the basis for what follows:

Definition 25: *Let* $\{\alpha_i\}_{i \in \mathbb{N}} \subset \mathbb{R}$ *be such that* $\gamma_i = \sum_{j=1}^{i}(\alpha_j + 1) > 0$ *for* $i \in \mathbb{N}$. *We call a point process* N *a* **generalized order statistic process** *w. r. t.* $\{\alpha_i\}_{i \in \mathbb{N}}$, *if, given* $N_t = n$ *for* $t > 0$ *and* $n \in \mathbb{N}$ *with* $P(N_t = n) > 0$, *then the conditional distribution of successive occurrence times* $T_1, \ldots, T_n$ *of* N *in* $[0, t]$ *is absolutely continuous w. r. t. the Lebesgue measure on* $\mathbb{R}^n$ *with density*

$$f_{T_1,\ldots,T_n|N_t=n}(t_1, \ldots, t_n) \;=\; t^{-\gamma_n} \prod_{i=1}^{n} (\gamma_i t_i^{\alpha_i}) \cdot \mathbb{1}_{K_n(t)}(t_1, \ldots, t_n). \tag{3.2}$$

The restrictions on $\{\gamma_i\}_{i \in \mathbb{N}}$ and thus $\{\alpha_i\}_{i \in \mathbb{N}}$ assure the nonnegativity and integrability of the conditional densities. For technical reasons we put $\gamma_0 = 0$.

Evidently, the above defined property generalizes Definition 19 (ii):

Example 26: For $\alpha_i = 0$, $i \in \mathbb{N}$, a GOS-process N satisfies the UOS-property and is thus a mixed Poisson process. □

Note that the existence of nontrivial GOS-processes, i. e. such that $P(T_1 < \infty) > 0$, w. r. t. a given parametrizing sequence is not evident a priori. On the contrary, there are cases which have to be excluded – see Section 4.3 for a counterexample. To get a feeling for the difficulty of the problem, note that we essentially deal with the question if and under which conditions to a given family of densities $\{f_{t,n}\}_{n\in\mathbb{N},t>0}$ concentrated on the cone $K_n(t) = \{(t_1,\ldots,t_n) \in \mathbb{R}^n \mid 0 < t_1 \leq \cdots \leq t_n \leq t\}$ there exists a point process N whose successive occurrence times $T_1, T_2, \ldots$ conditionally follow these densities, that is such that, given $N(t) = n$ for arbitrary $n \in \mathbb{N}$ and $t > 0$, the density of $T_1, \ldots, T_n$ equals $f_{t,n}$.

Note further, that in opposition to the behavior of mixed Poisson processes, there is as well the possibility that GOS-processes explode, i.e. that

$$P\left(\sup_{n\in\mathbb{N}} T_n < \infty\right) > 0.$$

It will turn out, that whether such a behavior occurs depends on the corresponding parametrizing sequence, compare Proposition 36 of Section 3.6. Explosion is especially present if the parameters grow too fast.

The outline for the rest of the chapter is as follows: Before we seek to study generalized order statistic processes in a comprehensive way we look in detail at the structure of the conditional densities given by equation (3.2) and establish a connection to generalized order statistics. This will be followed by Section 3.3 which provides basic analytical properties of GOS-processes. Section 3.4 characterizes GOS-processes in terms of their finite-dimensional distributions, Section 3.5 presents further distributional properties. The concluding Section 3.6 characterizes explosion of GOS-processes in terms of the corresponding parametrizing sequence. Altogether, we do not impose further assumptions on the parametrizing sequence $\{\alpha_i\}_{i\in\mathbb{N}}$ and present a collection of results which holds in this very general context. Contrariwise, Chapter 4 provides results departing from further restrictions.

Representations of the conditional distributions in terms of generalized order statistics

The following proposition clarifies the relation to the model of generalized order statistics, compare Section 2.1:

Proposition 27: *Let* $\alpha_1, \ldots, \alpha_n \in \mathbb{R}$ *be such that* $\gamma_i = \sum_{j=1}^{i}(\alpha_j + 1) > 0$ *for* $i = 1, \ldots, n$. *Further, for* $\tilde{\alpha} = (\alpha_n, \ldots, \alpha_1)$ *let* $X_{1:n}, \ldots, X_{n:n}$ *be generalized order statistics based on* $U[0,t]$. *Then,* $t - X_{n:n}, \ldots, t - X_{1:n}$ *admit the following joint density:*

$$f(t_1, \ldots, t_n) = t^{-\gamma_n} \prod_{i=1}^{n} \gamma_i t_i^{\alpha_i}, \qquad 0 < t_1 \leq \ldots \leq t_n \leq t.$$

Proof: For $n \in \mathbb{N}$, the density of $X_{1:n}, \ldots, X_{n:n}$, i.e. the density of generalized order statistics based on a uniform distribution on $[0,t]$, due to (2.1) equals

$$f_{X_{1:n},\ldots,X_{n:n}}(t_1,\ldots,t_n) = t^{-n} \prod_{i=1}^{n} \tilde{\gamma}_i \left(1 - \frac{t_i}{t}\right)^{\tilde{\alpha}_i},$$

where $\tilde{\gamma}_i = \sum_{j=i}^n (\tilde{\alpha}_j + 1)$, $i = 1, \ldots, n$, and we find

$$f_{t-X_{n:n},\ldots,t-X_{1:n}}(t_1,\ldots,t_n) = f_{X_{1:n},\ldots,X_{n:n}}(t - t_n, \ldots, t - t_1)$$

$$= t^{-n} \prod_{i=1}^n \tilde{\gamma}_i \left(1 - \frac{t - t_{n-i+1}}{t}\right)^{\tilde{\alpha}_i} = t^{-n} \prod_{i=1}^n \left(\sum_{j=i}^n (\tilde{\alpha}_j + 1)\right) \left(\frac{t_{n-i+1}}{t}\right)^{\tilde{\alpha}_i}$$

$$= t^{-n} \prod_{i=1}^n \left(\sum_{j=i}^n (\alpha_{n-j+1} + 1)\right) \left(\frac{t_{n-i+1}}{t}\right)^{\alpha_{n-i+1}}$$

$$= t^{-\gamma_n} \prod_{i=1}^n \gamma_i t_i^{\alpha_i}, \qquad 0 < t_1 \leq \cdots \leq t_n,$$

where the last equation holds since $\prod_{i=1}^n t_{n-i+1}^{\alpha_{n-i+1}} = \prod_{i=1}^n t_i^{\alpha_i}$ and

$$\prod_{i=1}^n \left(\sum_{j=i}^n (\alpha_{n-j+1} + 1)\right) = \prod_{i=1}^n \left(\sum_{j=1}^{n-i+1} (\alpha_j + 1)\right) = \prod_{i=1}^n \left(\sum_{j=1}^i (\alpha_j + 1)\right) = \prod_{i=1}^n \gamma_i. \quad \blacksquare$$

To summarize: For a generalized order statistic process N w.r.t. $\{\alpha_i\}_{i\in\mathbb{N}}$, given $N_t = n$, the times $t - T_n, \ldots, t - T_1$ are distributed like generalized order statistics $X_{1:n}, \ldots, X_{n:n}$ based on U$[0, t]$ and $\tilde{\alpha} = (\alpha_n, \ldots, \alpha_1)$, $n \in \mathbb{N}$, $t > 0$.

To analyze our densities given in (3.2), we could also refer to the model of so called dual order statistics, compare Definition 2. Since this model is directly derived from generalized order statistics, mainly inverting the increasing order, and does neither offer additional insight nor prevent us from a reparametrization, we stick to the more "classical" model of generalized order statistics.

Now, it is easy to find a constructive representation of the conditional densities of occurrence times in terms of Beta distributed random variables, cp. Appendix D for Beta distributions:

Formula (2.2) implies that there exist independent B$(\tilde{\gamma}_i, 1)$-distributed random variables $\tilde{B}_i$, $i = 1, \ldots, n$, such that

$$X_{i:n} = F^{-1}\left(1 - \prod_{j=1}^i \tilde{B}_j\right) = t\left(1 - \prod_{j=1}^i \tilde{B}_j\right), \qquad i = 1, \ldots, n,$$

holds for generalized order statistics $X_{1:n}, \ldots, X_{n:n}$ w.r.t. $\tilde{\alpha}$ and based on U$[0, t]$. Hence, according to Proposition 27, the occurrence times of a GOS-process N, given $N_t = n$, $t > 0$, $n \in \mathbb{N}$, can be represented by

$$T_i = t - X_{n-i+1:n} = t \prod_{j=1}^{n-i+1} \tilde{B}_j = t \prod_{j=i}^n \tilde{B}_{j-i+1}, \qquad i = 1, \ldots, n.$$

Since $\tilde{B}_{j-i+1}$ is $B(\tilde{\gamma}_{j-i+1}, 1)$-distributed and

$$\tilde{\gamma}_{j-i+1} = \sum_{l=j-i+1}^{n} (\tilde{\alpha}_l + 1) = \sum_{l=j-i+1}^{n} (\alpha_{n-l+1} + 1) = \sum_{l=1}^{n-j+i} (\alpha_l + 1) = \gamma_{n-j+i}$$

we finally obtain

$$T_i = t \prod_{j=i}^{n} B_j \qquad (3.3)$$

for independent $B(\gamma_i, 1)$-distributed random variables B_i, $i = 1, \ldots, n$.

This constructive representation of the conditional occurrence times corresponds to a stick-breaking mechanism:

Stick-breaking construction

Let $\alpha_1, \ldots, \alpha_n \in \mathbb{R}$ be such that $\gamma_i = \sum_{j=1}^{i}(\alpha_j + 1) > 0$, $i = 1, \ldots, n$. Further, consider a stick of length $t > 0$ and break off a random fraction E_1 where E_1 is Beta distributed $B(1, \gamma_n)$. Then, the remaining piece has length $R_1 = t - tE_1 = tE_1^c$ where $E_1^c = 1 - E_1$ is Beta distributed $B(\gamma_n, 1)$.

Therefrom again, break off an independent random fraction E_2 where E_2 is Beta distributed $B(1, \gamma_{n-1})$. Hence, we break off a piece of length $tE_1^c E_2$ and it remains the length $R_2 = tE_1^c - tE_1^c E_2 = tE_1^c E_2^c$ with $E_2^c = 1 - E_2$.

After n steps, we obtain a remaining piece of length $R_n = t \prod_{i=1}^{n} E_i^c$, where E_i^c is $B(\gamma_i, 1)$-distributed. Altogether, $R_n, \ldots, R_1$ are distributionally equal to the occurrence times $T_1, \ldots, T_n$ of a generalized order statistic process given $N_t = n$, $n \in \mathbb{N}$.

3.3 Analytical properties of generalized order statistic processes

This section presents two preliminary lemmas which provide basic analytical properties of generalized order statistic processes.

Lemma 28: *Let N be a generalized order statistic process w. r. t. $\{\alpha_i\}_{i \in \mathbb{N}}$ such that $P(N_t = n) > 0$ for some $n \in \mathbb{N}_0$ and $t > 0$. Then $P(N_s = m) > 0$ for $0 < s < t$ and $m \in \mathbb{N}_0$, $m \leq n$.*

Proof: The statement is evident for $n = 0$ since $\{N_t = 0\} \subset \{N_s = 0\}$ for $0 < s < t$. Else, let $P(N_t = n) > 0$ for some $n \in \mathbb{N}$ and $t > 0$. Then due to the generalized order statistic property we find for $0 < s < t$ and $m \leq n$

$$P(N_s = m) \geq P(N_s = m | N_t = n) P(N_t = n)$$

$$= P(T_1, \ldots, T_m \leq s < T_{m+1}, \ldots, T_n \leq t | N_t = n) P(N_t = n)$$

$$= \int \cdots \int_B \prod_{i=1}^{n} \gamma_i x_i^{\alpha_i} \, d(x_n \ldots x_1) P(N_t = n),$$

where

$$
B = \left\{ (x_1, \ldots, x_n) \in \mathbb{R}^n \; \middle| \;
\begin{array}{l}
0 < x_i \leq s, \quad i = 1, \ldots, m \\
s \leq x_i \leq t, \quad i = m+1, \ldots, n \\
x_i \leq x_{i+1}, \quad i = 1, \ldots, n-1
\end{array}
\right\},
$$

which exceeds 0 since $\ell^n(B) > 0$.   ∎

The next lemma presents a result which is the analog of Proposition 11 for GOS-processes. Put $p_{<\infty} = P(T_1 < \infty)$ and $p_\infty = P(T_1 = \infty)$.

Lemma 29: *For a generalized order statistic process N the distribution $P_{T_1, \ldots, T_n}$ of its occurrence times $T_1, \ldots, T_n$ can be decomposed as follows:*

$$
P_{T_1, \ldots, T_n} = P(T_1 < \infty) f_{T_1, \ldots, T_n} \odot \bar{\ell}^n + P(T_1 = \infty) \delta_{(\infty, \ldots, \infty)}, \qquad (3.4)
$$

where $f_{T_1, \ldots, T_n}$ is a probability density on $\mathbb{R}^n$, $n \in \mathbb{N}$.

Proof: Let N be such that $p_{<\infty} > 0$, otherwise the statement is trivial. Fix a $n \in \mathbb{N}$. For a set $A \in \bar{\mathcal{B}}^n$ we find

$$
P_{T_1, \ldots, T_n}(A) = p_{<\infty} P\big((T_1, \ldots, T_n) \in A | T_1 < \infty\big) + p_\infty P\big((T_1, .., T_n) \in A | T_1 = \infty\big).
$$

For the probability provided that at least one jump occurs we obtain

$$
P\big((T_1, \ldots, T_n) \in A | T_1 < \infty\big) = P\big((T_1, \ldots, T_n) \in A \cap \mathbb{R}^n | T_1 < \infty\big), \qquad A \in \bar{\mathcal{B}}^n,
$$

since $P(T_{n-1} < \infty, T_n = \infty) = 0$:

Let us assume the contrary, i.e., that $P(T_n < \infty, T_{n+1} = \infty) > 0$. As

$$
\{T_n < \infty, T_{n+1} = \infty\} = \cup_{i=1}^\infty \{T_n \leq i, T_{n+1} = \infty\}
$$

there exists $x > 0$ such that

$$
P(T_n \leq x, T_{n+1} = \infty) > 0. \qquad (3.5)
$$

Further

$$
P(T_n \leq x, T_{n+1} = \infty)
$$

$$
\leq P(T_n \leq x, N_y = n) = P(T_n \leq x | N_y = n) P(N_y = n)
$$

$$
\leq P(T_n \leq x | N_y = n) = \int_0^x \int_{s_1}^x \cdots \int_{s_{n-1}}^x y^{-\gamma_n} \prod_{i=1}^n \gamma_i s_i^{\alpha_i} ds_n \cdots ds_1 \qquad (3.6)
$$

holds for every $y \geq x$. Since expression (3.6) tends to 0 for y tending to ∞ this contradicts (3.5) and the assumption that $P(T_n < \infty, T_{n+1} = \infty) > 0$.

Now, let $B \in \mathcal{B}^n$ be an arbitrary set such that $\ell^n(B) = 0$. We find

$$
\begin{aligned}
0 &\leq P\big((T_1, \ldots, T_n) \in B | T_1 < \infty\big) \\
&\leq \sum_{x \in \mathbb{Q} \cap [0, \infty)} P\big((T_1, \ldots, T_n) \in B, N_x = n | T_1 < \infty\big) \\
&= \sum_{x \in \mathbb{Q} \cap [0, \infty)} P\big((T_1, \ldots, T_n) \in B | N_x = n\big) P(N_x = n | T_1 < \infty) = 0,
\end{aligned}
$$

because, given $N_x = n$, the distribution of $(T_1, \ldots, T_n)$ is absolutely continuous with respect to ℓ^n. Therefore, given $T_1 < \infty$, the distribution of $(T_1, \ldots, T_n)$ is absolutely continuous with respect to ℓ^n and admits a ℓ^n-density $f_{T_1, \ldots, T_n}$. As the distribution of $T_1, \ldots, T_n$, provided that a jump occurs, is concentrated on $\mathbb{R}^n$, $f_{T_1, \ldots, T_n}$ is a probability density on $\mathbb{R}^n$.

Altogether, we obtain

$$
\begin{aligned}
P_{T_1, \ldots, T_n}&(A) \\
&= p_{<\infty} P\big((T_1, .., T_n) \in A | T_1 < \infty\big) + p_{\infty} P\big((T_1, .., T_n) \in A | T_1 = \infty\big) \\
&= p_{<\infty} f_{T_1, \ldots, T_n} \odot \bar{\ell}^n(A) + p_{\infty} \delta_{(\infty, \ldots, \infty)}(A), \qquad A \in \mathcal{B}^n, \, n \in \mathbb{N},
\end{aligned}
$$

thereby $f_{T_1, \ldots, T_n}$ is to extend arbitrarily to $\bar{\mathbb{R}}^n$, compare the remarks ahead of Proposition 11. ∎

Remark 30: Note that this result especially means that a path of a generalized order statistic process N arriving in state 1 will increase to infinity with probability 1. In other words, N can be obtained as a mixture of two point processes one of which remains in the initial state 0 with probability 1 (with weight p_{∞}) whereas the other one, say $\tilde{N}$, is a process that increases to infinity almost surely (with weight $p_{<\infty}$). The distribution of the increasing point process can be obtained conditioning on $T_1 < \infty$:

$$
\begin{aligned}
P((T_1, \ldots, T_n) \in A | T_1 < \infty) &= \frac{P_{T_1, \ldots, T_n}(A \cap \{T_1 < \infty\})}{p_{<\infty}} \\
&= \frac{p_{<\infty} \int_{A \cap \{T_1 < \infty\}} f_{T_1, \ldots, T_n} d\bar{\ell}^n + p_{\infty} \delta_{(\infty, \ldots, \infty)}(A \cap \{T_1 < \infty\})}{p_{<\infty}} \\
&= \int_{A \cap \{T_1 < \infty\}} f_{T_1, \ldots, T_n} d\bar{\ell}^n = \int_A f_{T_1, \ldots, T_n} d\bar{\ell}^n, \qquad A \in \mathcal{B}^n,
\end{aligned}
$$

since $\bar{\ell}^n(A \cap \{T_1 = \infty\}) = 0$. Additionally we find

$$
P(T_1 \leq t_1, \ldots, T_n \leq t_n | N_t = n, T_1 < \infty) = P(T_1 \leq t_1, \ldots, T_n \leq t_n | N_t = n)
$$

for $n \in \mathbb{N}$ and $0 < t_1 \leq \ldots \leq t_n$ which is why N, given $T_1 < \infty$, and with it $\tilde{N}$ are GOS-processes w. r. t. the parametrizing sequence corresponding to N.

In the sequel, if we speak of (joint, conditional, etc.) densities of the occurrence times $T_1, T_2, \ldots$ of a GOS-process N we mean those corresponding to the point process $\tilde{N}$, which is such that at least one jump occurs almost surely.

Remark 31: Further consequences of Lemma 29 are the following:

(i) For all $n \in \mathbb{N}$ we have $P(T_n < \infty) = P(T_1 < \infty)$.

(ii) The function $t \mapsto P(N_t = n)$ is continuous on $(0, \infty)$ for arbitrary $n \in \mathbb{N}_0$.

3.4   Representation theorem

Let us gain further insight into the structure of GOS-processes deducing a characterization in terms of the basic structure of the (unconditional) densities associated to their successive occurrence times:

Theorem 32: *Let $\{\alpha_i\}_{i \in \mathbb{N}} \subset \mathbb{R}$ be a sequence and N a point process. Then the following statements are equivalent:*

(i) The process N is a generalized order statistic process w.r.t. $\{\alpha_i\}_{i \in \mathbb{N}}$.

(ii) There exists a family $\{\varphi_n\}_{n \in \mathbb{N}}$ of continuous functions $\varphi_n : (0, \infty) \to [0, \infty)$ such that, given $T_1 < \infty$, the successive occurrence times $T_1, \ldots, T_n$, $n \in \mathbb{N}$, follow the density

$$f_{T_1, \ldots, T_n}(t_1, \ldots, t_n) = \prod_{i=1}^{n} t_i^{\alpha_i} \cdot \varphi_n(t_n), \qquad 0 < t_1 \leq \cdots \leq t_n. \qquad (3.7)$$

Further, if (i) and (ii) hold we find $\gamma_i > 0$, $i \in \mathbb{N}$, and

$$P(T_1 < \infty)\varphi_n(t) = \frac{\prod_{i=1}^{n} \gamma_i}{t^{\gamma_n}} P(N_t = n), \qquad n \in \mathbb{N}, \, t > 0. \qquad (3.8)$$

Proof: If N is a trivial point process, i.e. $P(T_1 = \infty) = 1$, the statement is obvious. Without loss of generality (w.l.o.g.) let further be $P(T_1 < \infty) = 1$, compare Remark 30.

$(i) \Rightarrow (ii)$: Assume that N is a GOS-process w.r.t. $\{\alpha_i\}_{i \in \mathbb{N}}$. Due to Lemma 29 it remains to show that for $n \in \mathbb{N}$ there exists a continuous function φ_n with

$$f_{T_1, \ldots, T_n}(t_1, \ldots, t_n) = \prod_{i=1}^{n} t_i^{\alpha_i} \cdot \varphi_n(t_n), \qquad 0 < t_1 \leq \cdots \leq t_n.$$

Fix a $t > 0$ and let $0 < t_1 \leq \cdots \leq t_n \leq t$, $n \in \mathbb{N}$. Moreover, to begin with, assume that $P(N_t = n) > 0$ which also yields $P(N_s = n) > 0$ for $0 < s \leq t$ due to Lemma 28. Then we obtain

$$f_{T_1, \ldots, T_n | N_t = n}(t_1, \ldots, t_n) = t^{-\gamma_n} \prod_{i=1}^{n} \gamma_i t_i^{\alpha_i}, \qquad 0 < t_1 \leq \cdots \leq t_n \leq t,$$

because of the generalized order statistic property. On the other hand, direct calculation yields

$$P(T_1 \leq t_1, \ldots, T_n \leq t_n | N_t = n) = \frac{P(T_1 \leq t_1, \ldots, T_n \leq t_n, N_t = n)}{P(N_t = n)}$$

$$= \frac{\int_0^{t_1} \int_0^{t_2} \cdots \int_0^{t_n} \int_t^\infty f_{T_1, \ldots, T_{n+1}}(s_1, \ldots, s_{n+1}) \, ds_{n+1} \cdots ds_1}{P(N_t = n)}$$

$$= \int_0^{t_1} \int_0^{t_2} \cdots \int_0^{t_n} \frac{\int_t^\infty f_{T_1, \ldots, T_{n+1}}(s_1, \ldots, s_n, s_{n+1}) \, ds_{n+1}}{P(N_t = n)} \, ds_n \cdots ds_1, \qquad n \in \mathbb{N}.$$

Hence, for $n \in \mathbb{N}$, $t > 0$ and $t_1, \ldots, t_n$ with $0 \leq t_1 \leq \cdots \leq t_n \leq t$ the density $f_{T_1, \ldots, T_{n+1}}$ (or a version of it) satisfies

$$\frac{\int_t^\infty f_{T_1, \ldots, T_{n+1}}(t_1, \ldots, t_n, s_{n+1}) \, ds_{n+1}}{P(N_t = n)} = t^{-\gamma_n} \prod_{i=1}^n \gamma_i t_i^{\alpha_i}$$

respectively

$$\int_t^\infty f_{T_1, \ldots, T_{n+1}}(t_1, \ldots, t_n, s_{n+1}) \, ds_{n+1} = P(N_t = n) \, t^{-\gamma_n} \prod_{i=1}^n \gamma_i t_i^{\alpha_i}. \qquad (3.9)$$

If we choose especially $t = t_n$ we obtain for $f_{T_1, \ldots, T_n}$ (or a version of it)

$$f_{T_1, \ldots, T_n}(t_1, \ldots, t_n) = \int_{t_n}^\infty f_{T_1, \ldots, T_{n+1}}(t_1, \ldots, t_n, s_{n+1}) \, ds_{n+1}$$

$$= P(N_{t_n} = n) \, t_n^{-\gamma_n} \prod_{i=1}^n \gamma_i t_i^{\alpha_i}, \qquad 0 < t_1 \leq \cdots \leq t_n,$$

thus (3.7) holds for $\varphi_n(t) = \prod_{i=1}^n \gamma_i \cdot \frac{P(N_t = n)}{t^{\gamma_n}}$.

If otherwise $P(N_t = n) = 0$ for some $t > 0$ and $n \in \mathbb{N}$, we find $P(N_s = n) = 0$ for $s > t$ due to Lemma 28 which yields $P(T_n \geq t) = 0$ as $P(T_n = \infty) = P(T_1 = \infty) = 0$. This implies that $f_{T_1, \ldots, T_n}$ is concentrated on $K_n(t)$ or a subset of it and (3.7) holds for φ_n such that $\varphi_n(s) = 0$ for $s \geq t$. In particular, $\varphi_n(t) = 0$ yields (3.8).

Altogether, (3.8) holds for arbitrary $t > 0$ and $n \in \mathbb{N}$ and as $t \mapsto P(N_t = n)$ is a continuous mapping on $(0, \infty)$, cp. Remark 31 (ii), the function φ_n is continuous as well, $n \in \mathbb{N}$.

$(ii) \Rightarrow (i)$: Assume there exists a family $\{\varphi_n\}_{n \in \mathbb{N}}$ of continuous functions $\varphi_n : (0, \infty) \to [0, \infty)$ such that the occurrence times $T_1, \ldots, T_n$ follow the density

$$f_{T_1, \ldots, T_n}(t_1, \ldots, t_n) = \prod_{i=1}^n t_i^{\alpha_i} \cdot \varphi_n(t_n), \qquad n \in \mathbb{N}, \, 0 < t_1 \leq \cdots \leq t_n.$$

Firstly, we find that $\gamma_i > 0$ for $i \in \mathbb{N}$ because if for some $i \in \mathbb{N}$ we had $\gamma_i \leq 0$ then

$$\int_0^\infty \int_0^{s_{i+1}} \cdots \int_0^{s_2} f_{T_1,\ldots,T_{i+1}}(s_1,\ldots,s_{i+1})\, ds_1 \cdots ds_{i+1}$$

$$= \int_0^\infty \int_0^{s_{i+1}} \int_0^{s_i} \cdots \int_0^{s_2} \prod_{j=1}^{i-1} s_j^{\alpha_j}\, ds_1 \cdots ds_{i-1}\, s_i^{\alpha_i}\, ds_i\, s_{i+1}^{\alpha_{i+1}} \varphi_{i+1}(s_{i+1})\, ds_{i+1}$$

$$= \int_0^\infty \int_0^{s_{i+1}} \frac{s_i^{\gamma_i-1}}{\prod_{j=1}^{i-1} \gamma_j} s_i^{\alpha_i}\, ds_i\, s_{i+1}^{\alpha_{i+1}} \varphi_{i+1}(s_{i+1})\, ds_{i+1}$$

$$= \int_0^\infty \int_0^{s_{i+1}} \underbrace{\frac{s_i^{\gamma_i-1}}{\prod_{j=1}^{i-1} \gamma_j}\, ds_i}_{=\infty}\, s_{i+1}^{\alpha_{i+1}} \varphi_{i+1}(s_{i+1})\, ds_{i+1} \;=\; \infty,$$

which contradicts the assumption that $f_{T_1,\ldots,T_{i+1}}$ is a probability density. Furthermore, if $P(N_t = n) > 0$ for $n \in \mathbb{N}$, $t > 0$, then direct calculation yields for the conditional densities of successive occurrence times

$$P(T_1 \leq t_1,\ldots,T_n \leq t_n | N_t = n) = \frac{P(T_1 \leq t_1,\ldots,T_n \leq t_n, N_t = n)}{P(N_t = n)}$$

$$= \frac{P(T_1 \leq t_1,\ldots,T_n \leq t_n \leq t < T_{n+1})}{P(N_t = n)}$$

$$= \frac{\int_0^{t_1}\int_0^{t_2} \cdots \int_0^{t_n}\int_t^\infty \mathbb{1}_{K_n(t)}(s_1,..,s_n) \prod_{i=1}^{n+1} s_i^{\alpha_i} \cdot \varphi_{n+1}(s_{n+1})\, ds_{n+1}\cdots ds_1}{P(N_t = n)}$$

$$= \int_0^{t_1}\cdots\int_0^{t_n} \mathbb{1}_{K_n(t)}(s_1,..,s_n) \prod_{i=1}^n s_i^{\alpha_i} \underbrace{\frac{\int_t^\infty s_{n+1}^{\alpha_{n+1}} \varphi_{n+1}(s_{n+1})\, ds_{n+1}}{P(N_t = n)}}_{=I}\, ds_n \cdots ds_1,$$

where $n \in \mathbb{N}$ and $t_1,\ldots,t_n \in \mathbb{R}$. Hence the conditional density $f_{T_1,\ldots,T_n | N_t = n}$ exists and is almost surely equal to

$$f_{T_1,\ldots,T_n | N_t = n}(t_1,\ldots,t_n) = \mathbb{1}_{K_n(t)}(t_1,\ldots,t_n) \prod_{i=1}^n t_i^{\alpha_i} \cdot I$$

$$= \mathbb{1}_{K_n(t)}(t_1,\ldots,t_n) t^{-\gamma_n} \prod_{i=1}^n \gamma_i t_i^{\alpha_i},$$

where the last identity holds since I does not depend on $t_1,\ldots,t_n$ and is such that $P(T_1 \leq t,\ldots,T_n \leq t | N_t = n) = 1$. Moreover, N is a GOS-process w.r.t. $\{\alpha_i\}_{i\in\mathbb{N}}$.

Lastly, note that (ii) implies (3.8) by direct calculus, i.e.

$$
\begin{aligned}
P(N_t = n) &= P(T_n \leq t < T_{n+1}) \\
&= \int_0^t \int_{s_1}^t \cdots \int_{s_{n-1}}^t \int_t^\infty \prod_{i=1}^{n+1} s_i^{\alpha_i} \varphi_{n+1}(s_{n+1}) ds_{n+1} ds_n \cdots ds_1 \\
&= \int_0^t \int_{s_1}^t \cdots \int_{s_{n-1}}^t \prod_{i=1}^n s_i^{\alpha_i} ds_n \cdots ds_1 \int_t^\infty s_{n+1}^{\alpha_{n+1}} \varphi_{n+1}(s_{n+1}) ds_{n+1} \\
&= \frac{t^{\gamma_n}}{\prod_{i=1}^n \gamma_i} \varphi_n(t), \qquad n \in \mathbb{N},\ t > 0,
\end{aligned}
$$

where the last equation holds due to the projectivity of the densities. ∎

In particular, Theorem 32 states the following: If a point process N satisfies the generalized order statistic property w.r.t. $\{\alpha_i\}_{i\in\mathbb{N}}$, then the conditional density of $n \in \mathbb{N}$ successive occurrence times $T_1, \ldots, T_n$ provided that $T_1 < \infty$ is factorizable in the way that

$$
f_{T_1,\ldots,T_n}(t_1, \ldots, t_n) = \prod_{i=1}^n t_i^{\alpha_i} \cdot \varphi_n(t_n), \qquad n \in \mathbb{N},\ 0 < t_1 \leq \cdots \leq t_n,
$$

holds for some family $\{\varphi_n\}_{n\in\mathbb{N}}$ of functions.

Example 33: (continuation of Example 26) Let us return to the case $\alpha_i = 0$, $i \in \mathbb{N}$. Since a GOS-process N with respect to the given sequence $\{\alpha_i\}_{i\in\mathbb{N}}$ is a mixed Poisson process, there exists a distribution W on $(0, \infty)$ such that for $n \in \mathbb{N}$

$$
f_{T_1,\ldots,T_n}(t_1, \ldots, t_n) = \int_{(0,\infty)} \lambda^n e^{-\lambda t_n} \, dW(\lambda), \qquad 0 < t_1 \leq \cdots \leq t_n,
$$

compare Proposition 11. Hence, in this case

$$
\varphi_n(t) = \int_{(0,\infty)} \lambda^n e^{-\lambda t} \, dW(\lambda) = (-1)^n \hat{W}^{(n)}(t), \qquad n \in \mathbb{N},\ 0 < t,
$$

where $\hat{W}^{(n)}$ for $n \in \mathbb{N}$ denotes the n-th derivative of the Laplace transform $\hat{W}$ of W. □

Additionally, put

$$
\varphi_0(t) = \int_t^\infty \varphi_1(s) s^{\alpha_1} \, ds = \int_t^\infty f_{T_1}(s) ds = P(T_1 > t | T_1 < \infty), \qquad t \geq 0.
$$

As $\{f_{T_1,\ldots,T_n}\}_{n\in\mathbb{N}}$ are densities of a projective family of probability measures, the special structure given in Theorem 32 yields the following for $\{\varphi_n\}_{n\in\mathbb{N}_0}$:

$$
\int_t^\infty \varphi_{n+1}(s) s^{\alpha_{n+1}} \, ds = \varphi_n(t), \qquad n \in \mathbb{N}_0,\ t > 0. \tag{3.10}
$$

Equation (3.10) implies the differentiability of φ_n for $n \in \mathbb{N}_0$ in $(0, \infty)$. It can be equivalently expressed by

$$-\varphi_{n+1}(t)\, t^{\alpha_{n+1}} \;=\; \dot{\varphi}_n(t), \qquad n \in \mathbb{N}_0,\ t > 0, \tag{3.11}$$

$$\lim_{t \to \infty} \varphi_n(t) \;=\; 0, \qquad n \in \mathbb{N}_0.$$

Since the functions φ_n, $n \in \mathbb{N}_0$, are recursively related via equations (3.10) resp. (3.11), actually, the distribution of a GOS-process is completely determined by just any of the functions φ_n, e.g. φ_0, and the sequence $\{\alpha_i\}_{i \in \mathbb{N}}$ (and the probability $P(T_1 < \infty)$ that a jump occurs). This aspect will be studied in more detail in Chapter 5, where the notion of a generator for functions playing the role of φ_0 will be introduced and analyzed. We will so be able to transfer the question whether GOS-processes w. r. t. a certain sequence exist to the analytical question whether corresponding generators exist, as it is possible to construct GOS-processes based on such a generator playing the role of φ_0.

3.5 Basic distributional properties

We have just seen that the distribution of a GOS-process can be completely described by a family $\{\varphi_n\}_{n \in \mathbb{N}}$ of functions, the parametrizing sequence and the probability that a jump occurs. This chapter is mainly dedicated to deduce a formulary for the finite dimensional distributions of the process states, its occurrence and sojourn times in terms of $\{\varphi_n\}_{n \in \mathbb{N}}$ and the parametrizing sequence. Formulas concerning the distribution of sojourn times or states of the process turn out to be remarkably more complex than comparable ones concerning the occurrence times since the generalized order statistic property is a distributional property suited to the sequence of occurrence times.

The first section is dedicated to analyze GOS-processes undergoing monomial time transformations:

3.5.1 Time transformations

The class of GOS-processes is closed with respect to certain time transformations:

Proposition 34: *Let M be a point process and N a generalized order statistic process w. r. t. $\{\alpha_i\}_{i \in \mathbb{N}}$. Let further*

$$M(t) \;=\; N\left(d \cdot t^c\right), \qquad t \geq 0,$$

hold for some positive constants c, d. Then, M is a generalized order statistic process w. r. t. the sequence $\{c \cdot (\alpha_i + 1) - 1\}_{i \in \mathbb{N}}$.

Proof: Denote by T_i^M respectively T_i^N, $i \in \mathbb{N}$, the occurrence times of M and N, respectively. We find

$$d \cdot \left(T_i^M\right)^c = T_i^N, \qquad i \in \mathbb{N},$$

which yields the following for the joint densities of successive occurrence times of M, provided that at least one jump occurs:

$$f_{T_1^M,\ldots,T_n^M}(t_1,\ldots,t_n) = f_{T_1^N,\ldots,T_n^N}\left(d \cdot t_1^c,\ldots,d \cdot t_n^c\right) \prod_{i=1}^{n} \left(cdt_i^{c-1}\right)$$

$$= \prod_{i=1}^{n} (dt_i^c)^{\alpha_i} \cdot \varphi_n\left(d \cdot t_n^c\right) \cdot \prod_{i=1}^{n} \left(cdt_i^{c-1}\right) = \prod_{i=1}^{n} t_i^{\tilde{\alpha}_i} \cdot \tilde{\varphi}_n\left(t_n\right) \qquad (3.12)$$

for $n \in \mathbb{N}$, $0 < t_1 \leq \cdots \leq t_n$ and where $\tilde{\alpha}_i = c \cdot (\alpha_i + 1) - 1$, $i \in \mathbb{N}$, and $\tilde{\varphi}_n(t) = c^n d^{\sum_{i=1}^{n}(\alpha_i+1)} \varphi_n\left(d \cdot t^c\right)$ for $n \in \mathbb{N}$, $t > 0$. Theorem 32 completes the proof. ∎

3.5.2 Distribution of occurrence times

From now on and for the rest of the chapter let $\{\alpha_i\}_{i\in\mathbb{N}}$ be a real sequence, N a generalized order statistic process w.r.t. $\{\alpha_i\}_{i\in\mathbb{N}}$, $T_1, T_2, \ldots$ its occurrence times and $\{\varphi_n\}_{n\in\mathbb{N}}$ the corresponding family of functions such that

$$f_{T_1,\ldots,T_n}(t_1,\ldots,t_n) = \prod_{i=1}^{n} t_i^{\alpha_i} \varphi_n(t_n), \qquad n \in \mathbb{N}, 0 < t_1 \leq \cdots \leq t_n.$$

In the sense of Lemma 29 and the subsequent Remark 30, speaking of densities in connection with a GOS-process N, we mean the appropriate densities of the corresponding GOS-process $\tilde{N}$, which increases to infinity almost surely and whose distribution can be obtained conditioning on the event that at least one jump occurs. We obtain as density of the n-th occurrence time T_n

$$f_{T_n}(t) = \int_0^t \int_0^{s_{n-1}} \cdots \int_0^{s_2} \prod_{i=1}^{n-1} s_i^{\alpha_i} t^{\alpha_n} \varphi_n(t) \, ds_1 \cdots ds_{n-1}$$

$$= \frac{t^{\gamma_{n-1}}}{\prod_{i=1}^{n-1} \gamma_i} t^{\alpha_n} \varphi_n(t) = \frac{t^{\gamma_{n-1}}}{\prod_{i=1}^{n-1} \gamma_i} \varphi_n(t), \qquad n \in \mathbb{N}, t > 0. \qquad (3.13)$$

For two adjacent occurrence times T_{n-1} and T_n we obtain the following joint density:

$$f_{T_{n-1},T_n}(t_{n-1},t_n) = \int_0^{t_{n-1}} \int_0^{s_{n-2}} \cdots \int_0^{s_2} \prod_{i=1}^{n-2} s_i^{\alpha_i} \cdot t_{n-1}^{\alpha_{n-1}} t_n^{\alpha_n} \varphi_n(t_n) \, ds_1 \cdots ds_{n-2}$$

$$= \frac{t_{n-1}^{\gamma_{n-1}-1} t_n^{\alpha_n}}{\prod_{i=1}^{n-2} \gamma_i} \varphi_n(t_n), \qquad n \geq 2, 0 < t_{n-1} \leq t_n.$$

The conditional density of T_n given $T_1, \ldots, T_{n-1}$, $n \in \mathbb{N}$, equals

$$f_{T_n|T_1,\ldots,T_{n-1}}(t_n|t_1,\ldots,t_{n-1}) = \frac{f_{T_1,\ldots,T_n}(t_1,\ldots,t_n)}{f_{T_1,\ldots,T_{n-1}}(t_1,\ldots,t_{n-1})}$$

$$= \frac{\prod_{i=1}^{n} t_i^{\alpha_i} \varphi_n(t_n)}{\prod_{i=1}^{n-1} t_i^{\alpha_i} \varphi_{n-1}(t_{n-1})}$$

$$= t_n^{\alpha_n} \frac{\varphi_n(t_n)}{\varphi_{n-1}(t_{n-1})} \overset{(3.11)}{=} -\frac{\dot{\varphi}_{n-1}(t_n)}{\varphi_{n-1}(t_{n-1})}$$

$$= f_{T_n|T_{n-1}}(t_n|t_{n-1}),$$

for $0 < t_1 \leq \cdots \leq t_{n-1} \leq t_n$ such that $\varphi_{n-1}(t_{n-1}) > 0$. This formula implies that the sequence of occurrence times is a Markov chain. Besides we can observe that the conditional structure of the process, i. e., given the first $n-1$ occurrence times, resembles that of the original process. To be more precise, the density of $T_n, \ldots, T_{n+l}$ given $T_1, \ldots, T_{n-1}$ for $n \in \mathbb{N}$ and $l \in \mathbb{N}_0$ satisfies

$$f_{T_n,\ldots,T_{n+l}|T_1,\ldots,T_{n-1}}(t_n,\ldots,t_{n+l}|t_1,\ldots,t_{n-1}) = \frac{f_{T_1,\ldots,T_{n+l}}(t_1,\ldots,t_{n+l})}{f_{T_1,\ldots,T_{n-1}}(t_1,\ldots,t_{n-1})}$$

$$= \frac{\prod_{i=1}^{n+l} t_i^{\alpha_i} \varphi_{n+l}(t_{n+l})}{\prod_{i=1}^{n-1} t_i^{\alpha_i} \varphi_{n-1}(t_{n-1})} = \prod_{i=n}^{n+l} t_i^{\alpha_i} \varphi_{l+1}^{t_{n-1}}(t_{n+l}), \qquad (3.14)$$

for $0 < t_1 \leq \cdots \leq t_{n-1} \leq t_n \leq \cdots \leq t_{n+l}$ such that $\varphi_{n-1}(t_{n-1}) > 0$ and where $\varphi_{l+1}^{t_{n-1}}(t_{n+l}) = \frac{\varphi_{n+l}(t_{n+l})}{\varphi_{n-1}(t_{n-1})}$.

3.5.3 Normalized occurrence times and occurrence time ratios

Let us study the joint distribution of the normalized occurrence times $\frac{T_1}{T_n}, \ldots, \frac{T_{n-1}}{T_n}$, $n > 1$, given $T_1 < \infty$:

$$P\left(\frac{T_1}{T_n} \leq t_1, \ldots, \frac{T_{n-1}}{T_n} \leq t_{n-1} \middle| T_1 < \infty\right)$$

$$= P(T_1 \leq t_1 T_n, \ldots, T_{n-1} \leq t_{n-1} T_n | T_1 < \infty)$$

$$= \int_0^\infty \int_0^{t_1 s_n} \int_{s_1}^{t_2 s_n} \cdots \int_{s_{n-2}}^{t_{n-1} s_n} \prod_{i=1}^{n} s_i^{\alpha_i} \varphi_n(s_n) \, ds_{n-1} \cdots ds_1 ds_n$$

$$= \int_0^\infty \int_0^{t_1} \int_{s_1}^{t_2} \cdots \int_{s_{n-2}}^{t_{n-1}} \prod_{i=1}^{n-1} (s_n s_i)^{\alpha_i} s_n^{\alpha_n} \varphi_n(s_n) s_n^{n-1} \, ds_{n-1} \cdots ds_1 ds_n$$

$$
= \int_0^\infty s_n^{\gamma_n - 1} \varphi_n(s_n)\, ds_n \int_0^{t_1} \int_{s_1}^{t_2} \cdots \int_{s_{n-2}}^{t_{n-1}} \prod_{i=1}^{n-1} s_i^{\alpha_i}\, ds_{n-1} \cdots ds_1
$$

$$
= \prod_{i=1}^{n-1} \gamma_i \int_0^{t_1} \int_{s_1}^{t_2} \cdots \int_{s_{n-2}}^{t_{n-1}} \prod_{i=1}^{n-1} s_i^{\alpha_i}\, ds_{n-1} \cdots ds_1
$$

for $n \in \mathbb{N}$ and $0 < t_1 \le t_2 \le \cdots \le t_{n-1} \le 1$. The last equality holds since $\frac{s_n^{\gamma_n - 1}}{\prod_{i=1}^{n-1} \gamma_i} \varphi_n(s_n)$ is the density of T_n, compare formula (3.13). Thus the density of $\frac{T_1}{T_n}, \ldots, \frac{T_{n-1}}{T_n}$, provided that $T_1 < \infty$, is given by

$$
f_{\frac{T_1}{T_n}, \ldots, \frac{T_{n-1}}{T_n}}(t_1, \ldots, t_{n-1}) = \prod_{i=1}^{n-1} \gamma_i t_i^{\alpha_i}, \qquad 0 < t_1 \le \cdots \le t_{n-1} \le 1,\, n \in \mathbb{N},
$$

and we recover a distribution similar to that one defining the generalized order statistic property. Successive ratios of occurrence times $\frac{T_1}{T_2}, \frac{T_2}{T_3}, \ldots, \frac{T_{n-1}}{T_n}$ and T_n, $n \in \mathbb{N}$, follow the joint density

$$
f_{\frac{T_1}{T_2}, \frac{T_2}{T_3}, \ldots, \frac{T_{n-1}}{T_n}, T_n}(s_1, \ldots, s_n)
$$

$$
= f_{T_1, \ldots, T_n}(s_1 s_2 \cdots s_n, s_2 \cdots s_n, \ldots, s_n) \prod_{i=1}^{n} \prod_{j=i+1}^{n} s_j
$$

$$
= \prod_{i=1}^{n-1} \gamma_i s_i^{\gamma_i - 1} \cdot \frac{s_n^{\gamma_n - 1}}{\prod_{i=1}^{n-1} \gamma_i} \varphi_n(s_n) \tag{3.15}
$$

$$
= \prod_{i=1}^{n-1} f_{\frac{T_i}{T_{i+1}}}(s_i) \cdot f_{T_n}(s_n), \qquad 0 < s_i \le 1,\, i = 1, \ldots, n-1,\, s_n > 0.
$$

This remarkable property means that, provided that at least one jump occurs, the ratios $\frac{T_1}{T_2}, \frac{T_2}{T_3}, \ldots, \frac{T_{n-1}}{T_n}$ and T_n are independent and that $\frac{T_i}{T_{i+1}}$ is $\mathrm{B}(\gamma_i, 1)$-distributed, $i \in \mathbb{N}$. Again, it reflects an analogous "conditional" property, compare (3.3), where we observed that, given $N_t = n$ for $n \in \mathbb{N}$ and $t > 0$, successive occurrence time ratios $\frac{T_i}{T_{i+1}}$, $i = 1, \ldots, n-1$, are independently $\mathrm{B}(\gamma_i, 1)$-distributed.

3.5.4 Distribution of sojourn times

By a change of variables we obtain for the density of the first $n \in \mathbb{N}$ sojourn times $S_1, \ldots, S_n$, where $S_i = T_i - T_{i-1}$ for $i \in \mathbb{N}$ with $T_0 = 0$, the following:

$$
f_{S_1, \ldots, S_n}(s_1, \ldots, s_n) = f_{T_1, \ldots, T_n}\left(s_1, s_1 + s_2, \ldots, \sum_{i=1}^{n} s_i \right)
$$

$$
= \prod_{i=1}^{n} \left(\sum_{j=1}^{i} s_j \right)^{\alpha_i} \varphi_n\left(\sum_{j=1}^{n} s_j \right), \qquad s_1, \ldots, s_n > 0. \tag{3.16}
$$

3.5.5 Distribution of states

Theorem 32 yields

$$P(N_t = k) \;=\; p_{<\infty}\frac{t^{\gamma_k}}{\prod_{i=1}^{k}\gamma_i}\varphi_k(t), \qquad k \in \mathbb{N},\, t > 0. \qquad (3.17)$$

For $k = 0$ we find

$$P(N_t = 0) \;=\; p_{<\infty}\varphi_0(t) + p_\infty, \qquad t > 0, \qquad (3.18)$$

where $\varphi_0(t) \;=\; \int_t^\infty s^{\alpha_1}\varphi_1(s)ds$. For the rest of the section fix $n \in \mathbb{N}$, times $t_1,\ldots,t_n \in \mathbb{R}$ such that $0 = t_0 < t_1 \le t_2 \cdots \le t_n$ and states $k_1,\ldots,k_n \in \mathbb{N}_0$ such that $0 = k_0 \le k_1 \le \cdots \le k_n$ and $k_n > 0$. We find

$$P(N_{t_1} = k_1, N_{t_2} = k_2, \ldots, N_{t_n} = k_n)$$

$$= \; P(T_{k_1} \le t_1 < T_{k_1+1}, T_{k_2} \le t_2 < T_{k_2+1}, \ldots, T_{k_n} \le t_n < T_{k_n+1})$$

$$= \; p_{<\infty}\int\cdots\int_B \prod_{i=1}^{k_n+1} s_i^{\alpha_i}\varphi_{k_n+1}(s_{k_n+1})\, d(s_1,\ldots,s_{k_n+1}),$$

where

$$B \;=\; \left\{(s_1,\ldots,s_{k_n+1}) \in \mathbb{R}^{k_n+1} \;\middle|\; \begin{array}{ll} 0 < s_i \le s_{i+1}, & i = 1,\ldots,k_n \\ s_{k_i} \le t_i < s_{k_i+1}, & i = 1,\ldots,n \end{array}\right\}.$$

Denote by B_i, $i = 1,\ldots,n$, the canonical projection of B onto its $(k_{i-1}+1)$-, $\ldots$, k_i-th coordinates , i.e.

$$B_i = \{(s_{k_{i-1}+1},\ldots,s_{k_i}) \in \mathbb{R}^{k_i-k_{i-1}} | t_{i-1} < s_j \le s_{j+1} \le t_i, j = k_{i-1}+1,\ldots,k_i-1\},$$

and obtain

$$P(N_{t_1} = k_1, N_{t_2} = k_2, \ldots, N_{t_n} = k_n)$$

$$= p_{<\infty}\prod_{i=1}^{n}\int\cdots\int_{B_i}\prod_{j=k_{i-1}+1}^{k_i} s_j^{\alpha_j}\, d(s_{k_{i-1}+1},\ldots,s_{k_i})\cdot\int_{t_n}^{\infty} s_{k_n+1}^{\alpha_{k_n+1}}\varphi_{k_n+1}(s_{k_n+1})\, ds_{k_n+1}$$

$$= p_{<\infty}\prod_{i=1}^{n}\Psi_{k_i-k_{i-1}}\left(t_{i-1},t_i\,\middle|\,\alpha_{k_{i-1}+1},\ldots,\alpha_{k_i}\right)\varphi_{k_n}(t_n), \qquad (3.19)$$

where $\Psi_0(x,y) = 1$ and for $k \in \mathbb{N}$

$$\Psi_k(x,y|\alpha_1,\ldots,\alpha_k) \;=\; \int_x^y\int_{s_1}^y\cdots\int_{s_{k-1}}^y\prod_{j=1}^{k}s_j^{\alpha_j}ds_k\cdots ds_1, \qquad 0 < x \le y. \qquad (3.20)$$

Note that $\Psi_k(x,y|\alpha_1,\dots,\alpha_k)$ can be expressed in terms of the distribution of $U^d_{k:k}$ – the smallest of k dual generalized order statistics based on $U[0,1]$ and with parameters $\alpha_k,\alpha_{k-1},\dots,\alpha_1$, compare Definition 2 – because a change of variables shows

$$\int_x^y \int_{s_1}^y \cdots \int_{s_{k-1}}^y \prod_{j=1}^k s_j^{\alpha_j} \, ds_k \cdots ds_1 \;=\; \int_{\frac{x}{y}}^1 \int_{s_1}^1 \int_{s_2}^1 \cdots \int_{s_{k-1}}^1 \prod_{j=1}^k (s_j \cdot y)^{\alpha_j} \cdot y^k \, ds_k \cdots ds_1$$

$$=\; \frac{y^{\gamma_k}}{\prod_{j=1}^k \gamma_j} \int_{\frac{x}{y}}^1 \int_{s_1}^1 \int_{s_2}^1 \cdots \int_{s_{k-1}}^1 \prod_{j=1}^k \gamma_j s_j^{\alpha_j} \, ds_k \cdots ds_1$$

$$\overset{(2.4)}{=}\; \frac{y^{\gamma_k}}{\prod_{j=1}^k \gamma_j} P\left(U^d_{k:k} \geq \frac{x}{y} \right), \qquad 0 < x \leq y, \; k \in \mathbb{N}.$$

Equation (2.5) yields an alternative representation in terms of Meijer's G-function, compare Section A.3:

$$\Psi_k(x,y|\alpha_1,\dots,\alpha_k) \;=\; y^{\gamma_k} \int_{\frac{x}{y}}^1 G_k\left[y|\gamma_k,\gamma_{k-1},\dots,\gamma_1\right] dy, \qquad (3.21)$$

with $0 < x \leq y$, $k \in \mathbb{N}_0$. Finally, we obtain

$$P(N_{t_1} = k_1, N_{t_2} = k_2, \dots, N_{t_n} = k_n)$$

$$=\; p_{<\infty} \prod_{i=1}^n \left(t_i^{\gamma_{k_i} - \gamma_{k_{i-1}}} \int_{\frac{t_{i-1}}{t_i}}^1 G_{k_i - k_{i-1}}\left[y|\gamma_{k_i} - \gamma_{k_{i-1}}, \dots, \gamma_{k_{i-1}+1} - \gamma_{k_{i-1}}\right] dy \right) \varphi_{k_n}(t_n).$$

For the conditional probabilities $P(N(t_n) = k_n | N(t_1) = k_1, \dots, N(t_{n-1}) = k_{n-1})$ given the states in some past times we obtain

$$P(N_{t_n} = k_n | N_{t_1} = k_1, \dots, N_{t_{n-1}} = k_{n-1}) \;=\; \frac{P(N_{t_1} = k_1, N_{t_2} = k_2, \dots, N_{t_n} = k_n)}{P(N_{t_1} = k_1, \dots, N_{t_{n-1}} = k_{n-1})}$$

$$\overset{(3.19)}{=}\; \frac{p_{<\infty} \prod_{i=1}^n \Psi_{k_i - k_{i-1}}\left(t_{i-1}, t_i \,|\alpha_{k_{i-1}+1}, \dots, \alpha_{k_i}\right) \varphi_{k_n}(t_n)}{p_{<\infty} \prod_{i=1}^{n-1} \Psi_{k_i - k_{i-1}}\left(t_{i-1}, t_i \,|\alpha_{k_{i-1}+1}, \dots, \alpha_{k_i}\right) \varphi_{k_{n-1}}(t_{n-1})}$$

$$=\; \Psi_{k_n - k_{n-1}}\left(t_{n-1}, t_n \,|\alpha_{k_{n-1}+1}, \dots, \alpha_{k_n}\right) \frac{\varphi_{k_n}(t_n)}{\varphi_{k_{n-1}}(t_{n-1})} \qquad (3.22)$$

$$=\; P(N_{t_n} = k_n | N_{t_{n-1}} = k_{n-1}), \qquad k_{n-1} > 0,$$

if $P(N_{t_{n-1}} = k_{n-1}) > 0$. As further

$$P(N_{t_n} = k_n | N_{t_1} = 0, \dots, N_{t_{n-1}} = 0) \;=\; P(N_{t_n} = k_n | N_{t_{n-1}} = 0)$$

if $P(N_{t_{n-1}} = 0) > 0$, we find that N is a Markov process. For $P(N_t = j | N_s = 0)$ with $j \in \mathbb{N}$ and $0 < s \leq t$ we find

$$P(N_t = j | N_s = 0) \quad = \quad \frac{P(N_s = 0, N_t = j)}{P(N_s = 0)} \quad \overset{(3.18),(3.19)}{=} \quad \frac{p_{<\infty} \Psi_j(s,t|\alpha_1,\ldots,\alpha_j)\varphi_j(t)}{p_{<\infty}\varphi_0(s) + P(T_1 = \infty)}.$$

$$(3.23)$$

Additionally, note that the conditional probability $P(N_t = j | N_s = i)$ does in general depend on s and t and not exclusively on the time difference $t - s$, which is why a GOS-process, in general, is not homogeneous in the described sense. An exception constitute Poisson processes, where $\alpha_i = 0$ for $i \in \mathbb{N}$ and $\varphi_0(t) = e^{-\lambda t}$ for some $\lambda > 0$ and whose transition probabilities can be reduced to

$$P(N_t = j | N_s = i) \quad = \quad \frac{(\lambda(t-s))^{j-i}}{(j-i)!}e^{-\lambda(t-s)}, \qquad 0 < s \leq t,\, i,j \in \mathbb{N}_0,\, i \leq j.$$

Recurrence relation

The recurrence relation for $\{\varphi_n\}_{n \in \mathbb{N}}$ given by (3.11), that is

$$-t^{\alpha_n+1}\varphi_{n+1}(t) \quad = \quad \dot{\varphi}_n(t), \qquad n \in \mathbb{N}_0,\, t > 0,$$

results in a recurrence relation in terms of the probabilities $P(N_t = n)$, $t > 0$, $n \in \mathbb{N}_0$:

Proposition 35: *Let N be a generalized order statistic process w.r.t. $\{\alpha_i\}_{i \in \mathbb{N}}$. Then the probabilities $P(N_t = n)$ are differentiable and for $t > 0$ and $n \in \mathbb{N}_0$ we have*

$$\frac{d}{dt}P(N_t = n) \quad = \quad \frac{\gamma_n}{t}P(N_t = n) - \frac{\gamma_{n+1}}{t}P(N_t = n+1), \qquad (3.24)$$

where $\gamma_0 = 0$.

Proof: Let $t > 0$. Then we find

$$\frac{d}{dt}P(N_t = 0) \quad \overset{(3.18)}{=} \quad p_{<\infty}\dot{\varphi}_0(t) \quad \overset{(3.11)}{=} \quad -p_{<\infty}t^{\alpha_1}\varphi_1(t)$$

$$= \quad -p_{<\infty}\frac{\gamma_1}{t} \cdot \frac{t^{\gamma_1}}{\gamma_1}\varphi_1(t) \quad \overset{(3.17)}{=} \quad -\frac{\gamma_1}{t}P(N_t = 1).$$

For $n \in \mathbb{N}$ we have

$$\frac{d}{dt}P(N_t = n) \stackrel{(3.17)}{=} \frac{d}{dt}p_{<\infty}\frac{t^{\gamma_n}}{\prod_{i=1}^{n}\gamma_i}\varphi_n(t)$$

$$= p_{<\infty}\frac{t^{\gamma_n-1}}{\prod_{i=1}^{n-1}\gamma_i}\varphi_n(t) + p_{<\infty}\frac{t^{\gamma_n}}{\prod_{i=1}^{n}\gamma_i}\dot{\varphi}_n(t)$$

$$\stackrel{(3.11)}{=} p_{<\infty}\frac{t^{\gamma_n-1}}{\prod_{i=1}^{n-1}\gamma_i}\varphi_n(t) - p_{<\infty}\frac{t^{\gamma_{n+1}-1}}{\prod_{i=1}^{n}\gamma_i}\varphi_{n+1}(t)$$

$$\stackrel{(3.17)}{=} \frac{\gamma_n}{t}P(N_t = n) - \frac{\gamma_{n+1}}{t}P(N_t = n+1). \qquad \blacksquare$$

Remark the interesting similarity of (3.24) to formulas (2.9) resp. (2.10).

3.6 Explosion of generalized order statistic processes

The aim of this section is to characterize the exploding behavior of a GOS-process in terms of its parametrizing sequence. Firstly, due to Lemma 7 we find that $P\left(\sup_{n\in\mathbb{N}} T_n < \infty\right) > 0$ if and only if there exists a $t > 0$ such that

$$1 > \sum_{n=0}^{\infty} P(N_t = n) \stackrel{(3.17),(3.18)}{=} \sum_{n=0}^{\infty} p_{<\infty}\frac{t^{\gamma_n}}{\prod_{i=1}^{n}\gamma_i}\varphi_n(t) + p_{\infty}$$

which is equivalent to

$$\sum_{n=0}^{\infty} \frac{t^{\gamma_n}}{\prod_{i=1}^{n}\gamma_i}\varphi_n(t) < 1. \qquad (3.25)$$

However, it is hard to check whether (3.25) holds or not, especially if the functions φ_n cannot be specified explicitly. This inconvenience is resolved by the following proposition:

Proposition 36: *Let $\{\alpha_i\}_{i\in\mathbb{N}}$ be a real sequence and N a generalized order statistic process with respect to $\{\alpha_i\}_{i\in\mathbb{N}}$.*

(i) If $\displaystyle\sum_{i=0}^{\infty}\frac{1}{\gamma_i} = \infty$ then $P\left(\sup_{n\in\mathbb{N}} T_n = \infty\right) = 1$.

(ii) If $\displaystyle\sum_{i=0}^{\infty}\frac{1}{\gamma_i} < \infty$ then $P\left(\sup_{n\in\mathbb{N}} T_n < \infty | T_1 < \infty\right) = 1$.

Proof: W.l.o.g. $P(T_1 < \infty) = 1$. According to equation (3.15) we obtain

$$T_1 = \frac{T_1}{T_2} \cdot \frac{T_2}{T_3} \cdots \frac{T_n}{T_{n+1}} \cdot T_{n+1} = \prod_{i=1}^{n} B_i \cdot T_{n+1}, \qquad n \in \mathbb{N},$$

where B_i, $i \in \mathbb{N}$, are independent $B(\gamma_i, 1)$-distributed random variables such that additionally for $n \in \mathbb{N}$ the random variables $B_1, \ldots, B_n$ are independent of T_{n+1}. As $P(T_1 < \infty) = 1$ then $P\left(\sup_{n\in\mathbb{N}} T_n = \infty\right) = 1$ holds if and only if $\prod_{i=1}^{n} B_i$ tends to 0 for $n \to \infty$ almost surely. Since $\prod_{i=1}^{n} B_i$ is nonnegative this is satisfied if and only if

$$\prod_{i=1}^{n} E(B_i) = E\left(\prod_{i=1}^{n} B_i\right) \to 0 \qquad \text{for } n \to \infty, \tag{3.26}$$

due to dominated convergence. The expectation of a $B(\gamma_i, 1)$-distributed random variable equals $\frac{\gamma_i}{\gamma_i+1}$ that is why (3.26) is equivalent to

$$\lim_{n\to\infty} \prod_{i=1}^{n} \frac{\gamma_i}{\gamma_i + 1} = 0 \qquad \text{resp.} \qquad \lim_{n\to\infty} \prod_{i=1}^{n} \left(1 + \frac{1}{\gamma_i}\right) = \lim_{n\to\infty} \prod_{i=1}^{n} \frac{\gamma_i + 1}{\gamma_i} = \infty.$$

Since the product $\prod_{i=1}^{n} \left(1 + \frac{1}{\gamma_i}\right)$ diverges if and only if $\sum_{i=1}^{n} \frac{1}{\gamma_i}$ diverges, compare (Rainville, 1960, Thm. 3, p. 3), (i) follows.
As $\{\lim_{n\to\infty} \prod_{i=1}^{n} B_i = 0\}$ is a tail event and thus has probability 1 or 0, also (ii) follows and the proof is complete. ■

Due to the above result, explosion takes places for instance in the case of GOS-processes w. r. t. the sequence $\alpha_i = i$, $i \in \mathbb{N}$. This example will be studied later in Example 60 of Section 5.2 in a different context. In particular, we will simulate a corresponding sample path which will confirm the above characterization.

Remark 37: Note, that for GOS-processes with $P(T_1 < \infty) > 0$ and which do not explode, i. e. $P\left(\sup_{n\in\mathbb{N}} T_n < \infty\right) = 0$, we find that the corresponding functions φ_n, $n \in \mathbb{N}_0$ are strictly positive on $(0, \infty)$. Otherwise, if $\varphi_n(t) = 0$ for some $n \in \mathbb{N}_0$ and $t > 0$, the projectivity of the densities $f_{T_1,\ldots,T_n}$, compare (3.10) and (3.11), yields $\varphi_m(s) = 0$ for all $m \in \mathbb{N}_0$ and $s > t$. This again implies that the distributions of $T_1, \ldots, T_m$ are concentrated on $[0, t]$ for every $m \in \mathbb{N}$ and thus that infinitely many jumps occur in $[0, t]$ with probability 1, i. e. $P\left(\sup_{n\in\mathbb{N}} T_n \leq t\right) = 1$.

Remark 38: Note further, that a GOS-process w. r. t. a sequence $\{\alpha_i\}_{i\in\mathbb{N}}$ such that there exist $\alpha \in \mathbb{R}$ and $n^* \in \mathbb{N}$ with $\alpha_i = \alpha$ for $i > n^*$ does not explode almost surely according to Proposition 36.

Chapter 4

Generalized order statistic processes with respect to eventually constant sequences

Many results in the context of generalized order statistic processes can only be established imposing stronger assumptions on the parametrizing sequence $\{\alpha_i\}_{i\in\mathbb{N}}$. In particular, it is convenient to consider GOS-processes whose parametrizing sequences are eventually constant. Thereby, we call a sequence $\{\alpha_i\}_{i\in\mathbb{N}}$ eventually constant if there exists an index $n^* \in \mathbb{N}_0$ such that $\alpha_i = \alpha_{n^*+1}$ for $i > n^*$. The topic of the present chapter is thus to study such GOS-processes and their properties. Section 4.1 treats constant parametrizing sequences – a case where corresponding processes turn out to be time transformed mixed Poisson processes. In Section 4.2 we determine all GOS-processes with respect to eventually constant sequences, except for the case when the constant equals -1 which does not lead to consistent models of point processes, see Section 4.3. In Section 4.4, under further assumptions on $\{\alpha_i\}_{i\in\mathbb{N}}$, we present a way to construct these processes based on mixed Poisson processes. As a consequence we deduce an asymptotic result. Section 4.5 presents preliminary studies in order to transfer results achieved for eventually constant parametrizing sequences onto more general sequences.

4.1 Constant parametrizing sequences

The following result explains the nature of generalized order statistic processes if $\{\alpha_i\}_{i\in\mathbb{N}}$ is constant. It is a direct application of Theorem 22:

Proposition 39: *Let $\alpha > -1$ and N be a point process. Then there exists a mixed Poisson process M such that*

$$N_t = M(t^{\alpha+1}), \qquad t \geq 0,$$

if and only if N is a generalized order statistic process w. r. t. the constant sequence $\alpha_i = \alpha$, $i \in \mathbb{N}$.

Proof: Assume that $P(T_1 = \infty) < 1$, otherwise the statement is trivial.

If M is a mixed Poisson process and $N(t) = M(t^{\alpha+1})$, $t \geq 0$, we find the following conditional density for $n \in \mathbb{N}$ successive occurrence times of N:

$$f_{T_1^N, \ldots, T_n^N | N_t = n}(t_1, \ldots, t_n) = f_{(T_1^M)^{\frac{1}{\alpha+1}}, \ldots, (T_n^M)^{\frac{1}{\alpha+1}} | M(t^{\alpha+1}) = n}(t_1, \ldots, t_n)$$

$$= f_{T_1^M, \ldots, T_n^M | M(t^{\alpha+1}) = n}(t_1^{\alpha+1}, \ldots, t_n^{\alpha+1}) \prod_{i=1}^{n} (\alpha+1) t_i^{\alpha} \overset{\text{Thm. 18}}{=} \frac{n!(\alpha+1)^n}{t^{(\alpha+1)n}} \prod_{i=1}^{n} t_i^{\alpha},$$

where $0 < t_1 \leq \cdots \leq t_n \leq t$ and where T_i^N resp. T_i^M, $i \in \mathbb{N}$, denote the occurrence times of N resp. M. The above density is thus structured like in (3.2) with $\alpha_i = \alpha$ and $\gamma_i = i(\alpha+1)$, $i \in \mathbb{N}$. Thus N is a GOS-process.

Else, if N is a GOS-process w. r. t. $\alpha_i = \alpha$, $i \in \mathbb{N}$, the conditional densities (3.2) admit the following structure:

$$f_{T_1^N, \ldots, T_n^N | N_t = n}(t_1, \ldots, t_n) = n! \left(\frac{\alpha+1}{t} \right)^n \prod_{i=1}^{n} \left(\frac{t_i}{t} \right)^{\alpha} = n! \prod_{i=1}^{n} f_t(t_i)$$

for $0 < t_1 \leq \ldots \leq t_n \leq t$, where $f_t(s) = \frac{\alpha+1}{t} \left(\frac{s}{t} \right)^{\alpha}$ is the density of $F_t(s) = \left(\frac{s}{t} \right)^{\alpha+1}$, $0 \leq s \leq t$, $t > 0$. Thus, N has the order statistic property with respect to $\{F_t\}_{t>0}$ and due to Theorem 22 with $q(t) = ct^{\alpha+1}$ for some positive constant c, the point process N is a time transformed mixed Poisson process undergoing the time transformation $t \mapsto t^{\alpha+1}$. Remark that in order to apply Theorem 22 we need N to have non-exploding sample paths, i. e. to be such that $N(t) < \infty$ for $t > 0$. However, due to Proposition 36 and Remark 38 this does not impose a restriction here. ∎

Note that the above result implicitly states the existence of corresponding GOS-processes w. r. t. constant sequences.

4.2   Characterization theorem

The solution of the recursive system of differential equations (3.11) in the case of an eventually constant parametrizing sequence $\{\alpha_i\}_{i \in \mathbb{N}}$ is completely specified by the following proposition:

Theorem 40: *Let N be a point process and $\{\alpha_i\}_{i \in \mathbb{N}} \subset \mathbb{R}$ a sequence such that there exist $n^* \in \mathbb{N}$ and $\alpha > -1$ with $\alpha_i = \alpha > -1$ for $i > n^*$. Then the following statements are equivalent:*

(i) The process N is a generalized order statistic process w. r. t. $\{\alpha_i\}_{i \in \mathbb{N}}$.

(ii) There exists a probability distribution W on $(0, \infty)$ such that, provided that at least one jump occurs, the occurrence times $T_1, \ldots, T_n$ follow the ℓ^n-density

$$f_{T_1, \ldots, T_n}(t_1, \ldots, t_n) = \frac{\prod_{i=1}^{n} \gamma_i t_i^{\alpha_i}}{\Gamma \left(\frac{\gamma_n}{\alpha+1} + 1 \right)} \int_{(0,\infty)} \lambda^{\frac{\gamma_n}{\alpha+1}} e^{-\lambda t_n^{\alpha+1}} \, dW(\lambda) \qquad (4.1)$$

for $n \geq n^*$ *and* $0 < t_1 \leq \cdots \leq t_n$.

Remark again that this result implicitly includes that GOS-processes w. r. t. sequences eventually constant (exceeding -1) exist. Hence, the question posed below Definition 25, whether GOS-processes exist w. r. t. a given sequence, can be answered positively for a large class of parametrizing sequences.

Proof: W. l. o. g. let $P(T_1 < \infty) = 1$.

$(i) \Rightarrow (ii)$: Suppose that N is a GOS-process w. r. t. $\{\alpha_i\}_{i \in \mathbb{N}}$. Due to Theorem 32 we know that for $n \in \mathbb{N}$ there exists a function φ_n such that the density of $T_1, \ldots, T_n$ equals

$$f_{T_1,\ldots,T_n}(t_1,\ldots,t_n) = \prod_{i=1}^{n} t_i^{\alpha_i} \cdot \varphi_n(t_n), \qquad 0 < t_1 \leq \cdots \leq t_n.$$

It remains to show that there exists a distribution W on $(0,\infty)$ such that

$$\varphi_n(t) = \frac{\prod_{i=1}^{n} \gamma_i}{\Gamma\left(\frac{\gamma_n}{\alpha+1}+1\right)} \int_{(0,\infty)} \lambda^{\frac{\gamma_n}{\alpha+1}} e^{-\lambda t^{\alpha+1}} \, dW(\lambda), \qquad t > 0,$$

for $n \geq n^*$.

Therefore, consider the auxiliary functions $\psi_n(t) = \varphi_n\left(t^{\frac{1}{\alpha+1}}\right)$ for $t > 0$ and $n \geq n^*$. Actually ψ_{n^*} is the Laplace transform of some measure:

We show that ψ_{n^*} is completely monotone on $(0,\infty)$, compare Theorem B.3 in the Appendix. Rescaling the corresponding measure afterwards leads to the required probability measure W: With equation (3.11) we obtain

$$\frac{d}{dt}\psi_{n^*}(t) = \frac{d}{dt}\varphi_{n^*}\left(t^{\frac{1}{\alpha+1}}\right) = \frac{1}{\alpha+1} t^{\frac{1}{\alpha+1}-1} \dot{\varphi}_{n^*}\left(t^{\frac{1}{\alpha+1}}\right)$$

$$= -\frac{1}{\alpha+1} t^{\frac{1}{\alpha+1}-1} t^{\frac{\alpha}{\alpha+1}} \varphi_{n^*+1}\left(t^{\frac{1}{\alpha+1}}\right) = -(\alpha+1)^{-1}\psi_{n^*+1}(t), \qquad t > 0.$$

Repeating this procedure yields

$$\frac{d^k}{dt^k}\psi_{n^*}(t) = (-1)^k(\alpha+1)^{-k}\psi_{n^*+k}(t), \qquad k \in \mathbb{N}, t > 0.$$

Because $\psi_{n^*+k} \geq 0$ for $k \in \mathbb{N}$ we find

$$(-1)^k \frac{d^k}{dt^k}\psi_{n^*}(t) \geq 0, \qquad k \in \mathbb{N}, t > 0,$$

thus ψ_{n^*} is completely monotone on $(0,\infty)$ and there exists a measure V on $[0,\infty)$ such that

$$\psi_{n^*}(t) = \int_{[0,\infty)} e^{-\lambda t} \, dV(\lambda), \qquad t > 0.$$

For φ_{n^*} this implies

$$\varphi_{n^*}(t) = \psi_{n^*}(t^{\alpha+1}) = \int_{[0,\infty)} e^{-\lambda t^{\alpha+1}} \, dV(\lambda), \qquad t > 0.$$

Since $\varphi_{n^*}(t)$ approaches 0 for large t, the measure V cannot assign a positive weight to $\{0\}$, i.e., V is especially concentrated on $(0, \infty)$. Generally, V is no probability measure. Consider instead the measure W, absolutely continuous with respect to V such that

$$\frac{dW}{dV}(\lambda) = \frac{\Gamma\left(\frac{\gamma_{n^*}}{\alpha+1}+1\right)}{\prod_{i=1}^{n^*}\gamma_i}\lambda^{-\frac{\gamma_{n^*}}{\alpha+1}} \qquad \text{for } V\text{-almost all } \lambda.$$

Since

$$1 = \int_0^\infty \int_{s_1}^\infty \cdots \int_{s_{n^*-1}}^\infty \prod_{i=1}^{n^*} s_i^{\alpha_i}\varphi_{n^*}(s_{n^*})\,ds_{n^*}\cdots ds_1$$

$$= \int_0^\infty \int_{s_1}^\infty \cdots \int_{s_{n^*-1}}^\infty \prod_{i=1}^{n^*} s_i^{\alpha_i} \int_{(0,\infty)} e^{-\lambda s_{n^*}^{\alpha+1}}\,dV(\lambda)\,ds_{n^*}\cdots ds_1$$

$$= \int_{(0,\infty)} \underbrace{\int_0^\infty \int_0^{s_{n^*}} \cdots \int_0^{s_2} \prod_{i=1}^{n^*} s_i^{\alpha_i} e^{-\lambda s_{n^*}^{\alpha+1}}\,ds_1\cdots ds_{n^*}}\,dV(\lambda)$$

$$= \int_{(0,\infty)} \frac{\Gamma\left(\frac{\gamma_{n^*}}{\alpha+1}+1\right)}{\prod_{i=1}^{n^*}\gamma_i}\lambda^{-\frac{\gamma_{n^*}}{\alpha+1}} \qquad dV(\lambda) = \int_{(0,\infty)} dW(\lambda),$$

the measure W is indeed a probability measure. For φ_{n^*} we obtain

$$\varphi_{n^*}(t) = \int_{(0,\infty)} e^{-\lambda t^{\alpha+1}}\,dV(\lambda) = \frac{\prod_{i=1}^{n^*}\gamma_i}{\Gamma\left(\frac{\gamma_{n^*}}{\alpha+1}+1\right)}\int_{(0,\infty)} \lambda^{\frac{\gamma_{n^*}}{\alpha+1}}e^{-\lambda t^{\alpha+1}}\,dW(\lambda), \quad t>0.$$

By equation (3.11) we also have

$$\varphi_n(t) = \frac{(\alpha+1)^{n-n^*}\prod_{i=1}^{n^*}\gamma_i}{\Gamma\left(\frac{\gamma_{n^*}}{\alpha+1}+1\right)}\int_{(0,\infty)} \lambda^{\frac{\gamma_{n^*}}{\alpha+1}+(n-n^*)}e^{-\lambda t^{\alpha+1}}\,dW(\lambda)$$

$$= \frac{\prod_{i=1}^{n}\gamma_i}{\Gamma\left(\frac{\gamma_n}{\alpha+1}+1\right)}\int_{(0,\infty)} \lambda^{\frac{\gamma_n}{\alpha+1}}e^{-\lambda t^{\alpha+1}}\,dW(\lambda), \qquad t>0,$$

for $n \geq n^*$.

Theorem 32 proves that (ii) implies (i). ∎

Note again, that the assumptions of the above proposition explicitly exclude the case $\alpha = -1$ which generally would not disturb the conditions $\gamma_i > 0$ for $i \in \mathbb{N}$, yet leads to inconsistencies within the model, compare also Section 4.3.

Moreover, note that under the assumptions of Theorem 40 the distribution of the n^*-th occurrence time T_{n^*} is a mixture of generalized Γ-distributions such that

$$f_{T_{n^*}}(t) = \int_{(0,\infty)} \frac{(\alpha+1)\lambda^{\frac{\gamma_{n^*}}{\alpha+1}}}{\Gamma\left(\frac{\gamma_{n^*}}{\alpha+1}+1\right)}t^{\gamma_{n^*}-1}e^{-\lambda t^{\alpha+1}}\,dW(\lambda), \qquad t>0.$$

Example 41: Consider a delayed renewal process, that is a point process with independent sojourn times $S_i = T_i - T_{i-1}$, $i \in \mathbb{N}$, $T_0 = 0$, where $S_2, S_3, \ldots$ are identically distributed and where the first sojourn time S_1 possibly follows a different probability law. Let especially S_1 be Gamma distributed with the density

$$f_{S_1}(x) = \frac{\lambda^{\alpha_1 + 1}}{\Gamma(\alpha_1 + 1)} x^{\alpha_1} e^{-\lambda x}, \qquad x > 0,$$

where $\lambda > 0$, $\alpha_1 > -1$, and let $S_2, S_3, \ldots$ be exponentially distributed with intensity λ. Then we find the following joint density for successive occurrence times

$$
\begin{aligned}
f_{T_1, \ldots, T_n}(t_1, \ldots, t_n) &= f_{S_1, \ldots, S_n}(t_1, t_2 - t_1, \ldots, t_n - t_{n-1}) \\[2mm]
&= \frac{\lambda^{\alpha_1 + 1}}{\Gamma(\alpha_1 + 1)} t_1^{\alpha_1} e^{-\lambda t_1} \prod_{i=2}^{n} \lambda e^{-\lambda(t_i - t_{i-1})} \\[2mm]
&= \frac{\lambda^{\alpha_1 + n}}{\Gamma(\alpha_1 + 1)} t_1^{\alpha_1} e^{-\lambda t_n}, \qquad n \in \mathbb{N}, \ 0 < t_1 \le t_2 \le \ldots \le t_n,
\end{aligned}
$$

which admits a structure as given in equation (4.1) w. r. t. $\alpha_i = 0$ for $i = 2, 3, \ldots$ if we choose for W the distribution concentrated in a single λ. Hence, according to Theorem 40, the considered process satisfies the generalized order statistic property. □

Example 42: We want to illustrate the existence of GOS-processes whose coefficients α_i are smaller than -1 for some $i \in \{2, 3, \ldots\}$:
Choose for instance $\alpha_1 = 2$, $\alpha_2 = -2$ and $\alpha_i = 0$ for $i = 3, 4, \ldots$ Then we find $\gamma_1 = 3$, $\gamma_2 = 2$ and $\gamma_i = i$ for $i = 3, 4, \ldots$ According to Theorem 40 there exists a GOS-process with respect to the given sequence and its occurrence times follow the joint densities

$$
\begin{aligned}
f_{T_1, \ldots, T_n}(t_1, \ldots, t_n) &= \frac{\prod_{i=1}^{n} \gamma_i t_i^{\alpha_i}}{\Gamma\left(\frac{\gamma_n}{\alpha + 1} + 1\right)} \int_{(0,\infty)} \lambda^{\frac{\gamma_n}{\alpha + 1}} e^{-\lambda t_n^{\alpha + 1}} \, dW(\lambda) \\[2mm]
&= 3 \left(\frac{t_1}{t_2}\right)^2 \int_{(0,\infty)} \lambda^n e^{-\lambda t_n} \, dW(\lambda), \qquad n \ge 2,
\end{aligned}
$$

for some probability distribution W on $(0, \infty)$. This implies further

$$
\begin{aligned}
f_{T_1}(t) &= \int_t^{\infty} 3 \left(\frac{t}{s}\right)^2 \int_{(0,\infty)} \lambda^2 e^{-\lambda s} \, dW(\lambda) \, ds \\[2mm]
&= \int_{(0,\infty)} 3(\lambda t)^2 \left(\frac{e^{-\lambda t}}{t} - \lambda \Gamma(0, \lambda t)\right) dW(\lambda), \qquad t > 0,
\end{aligned}
$$

where $\Gamma(\cdot, \cdot)$ denotes the incomplete Gamma function, compare Section A.1. □

The final result of this section reveals further structural properties of GOS-processes w. r. t. eventually constant parametrizing sequences. Due to Proposition 39 we already know, that a GOS-process w. r. t. a constant sequence is a mixed Poisson process up to a time transformation. Thus, one might expect to recover this structure somehow after the n^*-th occurrence time of N. Actually, under the same assumptions as those of Theorem 40 we find:

Proposition 43: *Let N be a generalized order statistic process with respect to a sequence $\{\alpha_i\}_{i \in \mathbb{N}}$ such that $\alpha_i = \alpha$ for $i > n^*$, some $\alpha > -1$ and some $n^* \in \mathbb{N}$. Then, given $T_{n^*} = s$, the point process $\tilde{N}$, such that $\tilde{N}_t = N_{t+s} - N_s$ for $t \geq 0$, is a mixed Poisson process up to the time transformation $t \mapsto (t+s)^{\alpha+1} - s^{\alpha+1}$.*

Proof: Let $T_{n^*} = s$, $s \in (0, \infty)$, be given. Then the occurrence times $\tilde{T}_1, \tilde{T}_2, \ldots$ of $\tilde{N}$ verify $\tilde{T}_i = T_{n^*+i} - s$, $i \in \mathbb{N}$. Further, $P_{T_{n^*}}$-almost surely, for $s < \infty$, $P(\tilde{T}_i < \infty | T_{n^*} = s) = 1$ holds for $i \in \mathbb{N}$ and, given $T_{n^*} = s$, the occurrence times of $\tilde{N}$ are absolutely continuous w. r. t. the Lebesgue measure and follow the densities

$$f_{\tilde{T}_1,\ldots,\tilde{T}_n | T_{n^*} = s}(t_1, \ldots, t_n) \quad = \quad f_{T_{n^*+1},\ldots,T_{n^*+n} | T_{n^*} = s}(t_1 + s, \ldots, t_n + s)$$

$$\stackrel{(3.14)}{=} \quad \prod_{i=1}^{n} (t_i + s)^\alpha \varphi_n^s(t_n + s), \qquad 0 < t_1 \leq \cdots \leq t_n,$$

where $\varphi_n^s(t) = \frac{\varphi_{n^*+n}(t)}{\varphi_{n^*}(s)}$ for $t > 0$, $P_{T_{n^*}}$-almost surely. Apply Theorems 24 and 22: With $c(t) = (t+s)^{\alpha+1} - s^{\alpha+1}$ we have

$$f_{\tilde{T}_1,\ldots,\tilde{T}_n | T_{n^*} = s}(t_1, \ldots, t_n) \quad = \quad \prod_{i=1}^{n} c'(t_i)\, \psi(c(t_n)), \qquad n \in \mathbb{N}, 0 \leq t_1 \leq \cdots \leq t_n,$$

for ψ appropriately defined, which implies that $\tilde{N}$ satisfies the order statistic property w. r. t. $\{F_t\}_{t \geq 0}$, where $F_t(s) = \frac{c(s)}{c(t)}$, $0 \leq s \leq t$. Therefore, $\tilde{N}$ is a mixed Poisson process up to the time transformation $t \mapsto c(t)$. ∎

4.3   Nonexistence example

Theorem 40 and Proposition 43 only refer to GOS-processes w. r. t. eventually constant sequences with a constant exceeding -1. It turns out that there are no nontrivial processes satisfying a generalized order statistic property with respect to a parametrizing sequence eventually equal to -1, compare also the discussion below Example 26:

Proposition 44: *A nontrivial generalized order statistic process N w. r. t. parameters $\{\alpha_i\}_{i \in \mathbb{N}}$ such that $\alpha_i = -1$ for $i > n^*$ and some $n^* \in \mathbb{N}$ does not exist.*

Proof: Let $\{\alpha_i\}_{i \in \mathbb{N}}$ be such that $\alpha_i = -1$ for $i > n^*$ and some $n^* \in \mathbb{N}$, further such that $\gamma_i > 0$, $i \in \mathbb{N}$. Thus, all the densities in Definition 25 are properly defined. Assume that N is a GOS-process w.r.t. the given sequence. W.l.o.g. let $P(T_1 < \infty) = 1$, compare Remark 30. Moreover, note that N does not explode according to Remark 38 and that due to Remark 37 we have $P(N_t = n) > 0$ for $n \in \mathbb{N}_0$ and $t > 0$.

The idea of the first part of the proof is to eliminate $\alpha_1, \ldots, \alpha_{n^*}$ by conditioning on the first occurrence times $T_1, \ldots, T_{n^*}$ corresponding to these first n^* parameters possibly differing from -1. We will then show that, given $T_1, \ldots, T_{n^*}$, the remaining process is distributed like a time transformed mixed Poisson process.

For $n \geq n^*$, let $N_t = n$ be given and let $T_1, \ldots, T_n$ be the first n occurrence times in $[0, t]$ with the joint conditional density, compare (3.2),

$$f_{T_1, \ldots, T_n | N_t = n}(t_1, \ldots, t_n) = \frac{\prod_{i=1}^{n^*-1} \gamma_i \cdot \gamma_{n^*}^{n-n^*+1}}{t^{\gamma_{n^*}}} \cdot \frac{\prod_{i=1}^{n^*} t_i^{\alpha_i}}{t_{n^*+1} \cdots t_n}, \quad 0 < t_1 \leq \ldots \leq t_n \leq t.$$

Then $T_1, \ldots, T_{n^*}$, given $N_t = n$, follow the density

$$f_{T_1, \ldots, T_{n^*} | N_t = n}(t_1, \ldots, t_{n^*})$$

$$= \int_{t_{n^*}}^t \int_{t_{n^*+1}}^t \cdots \int_{t_{n-1}}^t \frac{\prod_{i=1}^{n^*-1} \gamma_i \cdot \gamma_{n^*}^{n-n^*+1}}{t^{\gamma_{n^*}}} \cdot \frac{\prod_{i=1}^{n^*} t_i^{\alpha_i}}{t_{n^*+1} \cdots t_n} \, dt_n dt_{n-1} \cdots dt_{n^*+1}$$

$$= \frac{\prod_{i=1}^{n^*-1} \gamma_i \cdot \gamma_{n^*}^{n-n^*+1} \cdot \prod_{i=1}^{n^*} t_i^{\alpha_i}}{t^{\gamma_{n^*}} (n-n^*)!} \cdot \left(\ln \left(\frac{t}{t_{n^*}} \right) \right)^{n-n^*}, \quad 0 < t_1 \leq \cdots \leq t_{n^*} \leq t.$$

Hence, the conditional density of $T_{n^*}, \ldots, T_n$ given $T_1, \ldots, T_{n^*}$ and $N_t = n$ is

$$f_{T_{n^*+1}, \ldots, T_n | T_1 = t_1, \ldots, T_{n^*} = t_{n^*}, N_t = n}(t_{n^*+1}, \ldots, t_n) = \frac{f_{T_1, \ldots, T_n | N_t = n}(t_1, \ldots, t_n)}{f_{T_1, \ldots, T_{n^*} | N_t = n}(t_1, \ldots, t_{n^*})}$$

$$= \frac{\dfrac{\prod_{i=1}^{n^*-1} \gamma_i \cdot \gamma_{n^*}^{n-n^*+1}}{t^{\gamma_{n^*}}} \cdot \dfrac{\prod_{i=1}^{n^*} t_i^{\alpha_i}}{t_{n^*+1} \cdots t_n}}{\dfrac{\prod_{i=1}^{n^*-1} \gamma_i \cdot \gamma_{n^*}^{n-n^*+1} \cdot \prod_{i=1}^{n^*} t_i^{\alpha_i}}{t^{\gamma_{n^*}} (n-n^*)!} \cdot \left(\ln \left(\dfrac{t}{t_{n^*}} \right) \right)^{n-n^*}}$$

$$= \frac{(n-n^*)!}{\left(\ln \left(\dfrac{t}{t_{n^*}} \right) \right)^{n-n^*} t_{n^*+1} \cdots t_n}, \quad 0 < t_1 \leq t_2 \leq \cdots \leq t_n \leq t.$$

This equals the density of $n - n^*$ order statistics based on the continuous distribution function

$$F(x) = \frac{\ln x - \ln t_{n^*}}{\ln t - \ln t_{n^*}}, \quad t_{n^*} \leq x \leq t,$$

compare equation (2.3). Consider the point process $\tilde{N}$ defined by

$$\tilde{N}_t = N(t + T_{n^*}) - 1, \qquad t \geq 0.$$

For the occurrence times $\tilde{T}_i$ of $\tilde{N}$ we obtain

$$\tilde{T}_i = T_{i+n^*} - T_{n^*}, \qquad i \in \mathbb{N}.$$

Furthermore, provided that $T_{n^*} = t_{n^*}$, $t_{n^*} > 0$, and $\tilde{N}_t = n$, $n \in \mathbb{N}$, the occurrence times $\tilde{T}_1, \ldots, \tilde{T}_n$ are distributed like ordinary order statistics based on the distribution

$$\tilde{F}(x) = \frac{\ln(x + t_{n^*}) - \ln t_{n^*}}{\ln(t + t_{n^*}) - \ln t_{n^*}}, \qquad 0 \leq x \leq t.$$

According to Theorem 22 with $q(t) = c(\ln(t + t_{n^*}) - \ln t_{n^*})$ for some positive constant c, the process $\tilde{N}$ given $T_{n^*} = t_{n^*}$ can be represented as

$$\tilde{N}_t = M(\ln(t + t_{n^*}) - \ln t_{n^*}), \qquad t \geq 0, \tag{4.2}$$

where M is a mixed Poisson process; for N we have

$$N_t = M(\ln t - \ln t_{n^*}) + 1, \qquad t \geq t_{n^*}. \tag{4.3}$$

By $V_{t_{n^*}}$ resp. $\hat{V}_{t_{n^*}}$ we denote the corresponding mixing distribution on $[0, \infty)$ resp. its Laplace transform, which possibly depend on t_{n^*}.

In the second part of the proof we attempt to determine the distribution of the n^*-th occurrence time T_{n^*}: Consider for $0 \leq x \leq t$ the probability

$$P(T_{n^*} \leq x | N_t = n^*).$$

The generalized order statistic property of N yields

$$P(T_{n^*} \leq x | N_t = n^*) = \left(\frac{x}{t}\right)^{\gamma_{n^*}}, \qquad 0 \leq x \leq t.$$

On the other hand, according to (4.3) we find

$$P(T_{n^*} \leq x | N_t = n^*) = \frac{P(T_{n^*} \leq x, N_t = n^*)}{P(N_t = n^*)}$$

$$= \frac{\int_0^x P(N_t = n^* | T_{n^*} = s) f_{T_{n^*}}(s) \, ds}{\int_0^t P(N_t = n^* | T_{n^*} = s) f_{T_{n^*}}(s) \, ds}$$

$$\stackrel{(4.3)}{=} \frac{\int_0^x \hat{V}_s(\ln t - \ln s) f_{T_{n^*}}(s) \, ds}{\int_0^t \hat{V}_s(\ln t - \ln s) f_{T_{n^*}}(s) \, ds}, \qquad 0 \leq x \leq t,$$

where $f_{T_{n^*}}$ denotes the density of T_{n^*}. Derivating with respect to x implies

$$\gamma_{n^*} \cdot \frac{x^{\gamma_{n^*}-1}}{t^{\gamma_{n^*}}} = \frac{\hat{V}_x(\ln t - \ln x) f_{T_{n^*}}(x)}{\int_0^t \hat{V}_s(\ln t - \ln s) f_{T_{n^*}}(s) \, ds}, \qquad 0 < x \leq t.$$

We obtain

$$\int_0^t \hat{V}_s(\ln t - \ln s) f_{T_{n^*}}(s)\, ds \;=\; \frac{\hat{V}_x(\ln t - \ln x) f_{T_{n^*}}(x)}{\gamma_{n^*} \cdot x^{\gamma_{n^*}-1}} \cdot t^{\gamma_{n^*}}$$

and since the left side does not depend on x we have

$$\hat{V}_x(\ln t - \ln x) f_{T_{n^*}}(x) \;=\; x^{\gamma_{n^*}-1} \cdot c(t)$$

or equivalently

$$f_{T_{n^*}}(x) \;=\; \frac{x^{\gamma_{n^*}-1} \cdot c(t)}{\hat{V}_x(\ln t - \ln x)}, \qquad 0 < x \le t, \tag{4.4}$$

where c is a positive real function. This representation of $f_{T_{n^*}}$ still depends on a parameter t. Our next objective is to eliminate t and to study the structure of the family of mixing distributions $\{V_s\}_{s\ge 0}$:

Consider two different times x, y such that $0 < x \le y \le t$. Then (4.4) implies

$$\hat{V}_y(\ln t - \ln y) \;=\; \hat{V}_x(\ln t - \ln x) \cdot \underbrace{\left(\frac{y}{x}\right)^{\gamma_{n^*}-1} \frac{f_{T_{n^*}}(x)}{f_{T_{n^*}}(y)}}_{=k(x,y)}$$

$$=\; \int_{[0,\infty)} e^{-\lambda(\ln t - \ln x)} k(x,y)\, dV_x(\lambda)$$

$$=\; \int_{[0,\infty)} e^{-\lambda(\ln t - \ln y)} \cdot e^{-\lambda(\ln y - \ln x)} \cdot k(x,y)\, dV_x(\lambda).$$

Since V_y is a probability distribution, we obtain for $0 < x \le y$

$$1 \;=\; \hat{V}_y(0) \;=\; \int_{[0,\infty)} e^{-\lambda(\ln y - \ln x)} \cdot k(x,y)\, dV_x(\lambda) \;=\; k(x,y) \cdot \hat{V}_x(\ln y - \ln x).$$

Thus, for two Laplace transforms $\hat{V}_x$ and $\hat{V}_y$ with $0 < x \le y \le t$ we find

$$\hat{V}_x(\ln t - \ln x) \;=\; \hat{V}_y(\ln t - \ln y) \cdot \hat{V}_x(\ln y - \ln x). \tag{4.5}$$

Lemma 45: *For*

$$G(x,t) = \hat{V}_{e^x}(t-x), \qquad 0 < x \le t,$$

we have

$$G(x,t) = \frac{\hat{W}(t)}{\hat{W}(x)},$$

where $\hat{W} : (0,\infty) \to (0,\infty)$ is a completely monotone function.

Proof: Note that G is positive as $\hat{V}_t$ is for every $t > 0$. Equation (4.5) implies

$$G(x,y) = \frac{G(x,t)}{G(y,t)}, \qquad 0 < x \leq y \leq t. \tag{4.6}$$

Partial derivation with respect to t yields

$$0 = \frac{G_t(x,t)G(y,t) - G(x,t)G_t(y,t)}{G(y,t)^2}$$

which results in

$$\frac{G_t(x,t)}{G(x,t)} = \frac{G_t(y,t)}{G(y,t)} = c(t), \qquad 0 < x \leq y \leq t.$$

Solving the differential equation $G_t(\cdot,t) = c(t)G(\cdot,t)$, we obtain the following representation

$$G(x,t) = \frac{\hat{W}_1(t)}{\hat{W}_2(x)}$$

for some positive functions $\hat{W}_1, \hat{W}_2$. In addition, (4.6) implies that $\hat{W}_1$ and $\hat{W}_2$ coincide. Put $\hat{W} = \hat{W}_1 = \hat{W}_2$. Since $G(x,t)$ is completely monotone in t for arbitrary $x > 0$ so is $\hat{W}$ and the proof of the lemma is complete. ∎

Due to Lemma 45 there exists a completely monotone function $\hat{W}$ on $(0,\infty)$ such that

$$\hat{V}_x(\ln t - \ln x) = \frac{\hat{W}(\ln t)}{\hat{W}(\ln x)}, \qquad 0 < x \leq t. \tag{4.7}$$

According to Theorem B.3, $\hat{W}$ is the Laplace transform of some measure W on $[0,\infty)$, and by equations (4.4) and (4.7) we obtain

$$f_{T_{n^*}}(x) = \frac{x^{\gamma_{n^*}-1}c(t)\hat{W}(\ln x)}{\hat{W}(\ln t)}, \qquad 0 < x \leq t.$$

Since for fixed $x > 0$ this holds for arbitrary $t \geq x$ the quotient $\frac{c(t)}{\hat{W}(\ln t)}$ has to be constant c and we obtain

$$f_{T_{n^*}}(x) = x^{\gamma_{n^*}-1}\hat{W}(\ln x)\cdot c, \qquad x > 0.$$

But Fubini's theorem yields

$$\int_0^\infty x^{\gamma_{n^*}-1}\hat{W}(\ln x)\,dx = \int_{[0,\infty)}\int_0^\infty x^{\gamma_{n^*}-1-\lambda}\,dx\,dW(\lambda) = \infty,$$

thus the above function $f_{T_{n^*}}$ cannot be a probability density which contradicts our assumptions, i.e. a GOS-process w.r.t. a sequence such that $\alpha_i = -1$ for $i > n^*$ cannot exist. ∎

4.4 A decomposition in terms of mixed Poisson processes

The aim of this section is to deduce a decomposition of GOS-processes w. r. t. eventually constant parametrizing sequences into mixed Poisson processes undergoing a time transformation and a deletion of certain process points.

To motivate our approach consider Example 41 of a delayed renewal process whose first occurrence time T_1 is $\Gamma(\lambda, \alpha_1 + 1)$-distributed with parameters $\lambda > 0$ and $\alpha_1 \in \mathbb{N}_0$, and whose sojourn times $S_i = T_i - T_{i-1}$, $i > 1$, are independent and identically $\text{Exp}(\lambda)$-distributed. We have seen that this process verifies the generalized order statistic property with respect to a sequence with $\alpha_i = 0$ for $i > 1$. Since a $\Gamma(\lambda, \alpha_1 + 1)$-distributed random variable can be obtained as the sum of $\alpha_1 + 1$ independent identically $\text{Exp}(\lambda)$-distributed random variables, the considered delayed renewal process can be modelled as a Poisson process with intensity λ whose first α_1 occurrence times have been deleted, that is $N_t = \max\{N_t^\lambda - \alpha_1, 0\}$ for $t \geq 0$, where N denotes the delayed renewal process and N^λ a Poisson process with intensity λ.

This idea is generalized in the subsequent Theorem 46. For a sequence $\{\tilde{\alpha}_i\}_{i \in \mathbb{N}}$ put $\tilde{\gamma}_i = \sum_{j=1}^{i}(\tilde{\alpha}_j + 1)$ for $i \in \mathbb{N}_0$.

Theorem 46: *Let N be a generalized order statistic process with respect to a sequence such that $\alpha_1, \ldots, \alpha_{n^*} \in \mathbb{N}$ for some $n^* \in \mathbb{N}$, $\alpha_i = 0$ for $i > n^*$, i.e. there exists some distribution W on $(0, \infty)$ such that its occurrence times, given $T_1 < \infty$, follow the densities*

$$f_{T_1,\ldots,T_n}(t_1, \ldots, t_n) = \frac{\prod_{i=1}^{n} \gamma_i t_i^{\alpha_i}}{\Gamma(\gamma_n + 1)} \int_{(0,\infty)} \lambda^{\gamma_n} e^{-\lambda t_n} \, dW(\lambda)$$

for $n \geq n^*$, $0 < t_1 \leq \cdots \leq t_n$.
Then, for the distribution P_N of N we obtain

$$P_N(A) = \sum_{\tilde{\alpha} \in I} w\begin{pmatrix} \alpha_1, \ldots, \alpha_{n^*} \\ \tilde{\alpha}_1, \ldots, \tilde{\alpha}_{n^*} \end{pmatrix} \cdot P_{\tilde{\alpha}}^V(A), \qquad A \in \mathcal{H}(\mathcal{N}),$$

with a (finite) index set

$$I = I(\alpha_1, \ldots, \alpha_{n^*}) = \left\{ \tilde{\alpha} \in \mathbb{N}_0^{n^*} \,\middle|\, \tilde{\gamma}_i \geq \gamma_i, \, i = 1, \ldots, n^* - 1; \, \tilde{\gamma}_{n^*} = \gamma_{n^*} \right\}$$

and weights

$$w\begin{pmatrix} \alpha_1, \ldots, \alpha_{n^*} \\ \tilde{\alpha}_1, \ldots, \tilde{\alpha}_{n^*} \end{pmatrix} = \frac{\prod_{i=1}^{n^*} \tilde{\alpha}_i! \gamma_i}{\Gamma(\gamma_{n^*} + 1)} \prod_{i=2}^{n^*} \binom{\tilde{\gamma}_i - \gamma_{i-1} - 1}{\tilde{\alpha}_i} \tag{4.8}$$

and where $P_{\tilde{\alpha}}^V$ denotes the distribution of a mixed Poisson process with mixing distribution V after deleting the points $T_{\tilde{\gamma}_{i-1}+1}, \ldots, T_{\tilde{\gamma}_i-1}$ for $\tilde{\gamma}_{i-1} + 1 \leq \tilde{\gamma}_i - 1$ and $i = 1, \ldots, n^*$. The mixing distribution V is concentrated on $[0, \infty)$ and such that $V(\{0\}) = P(T_1 = \infty)$ and $V((a, b)) = P(T_1 < \infty) W((a, b))$ for $0 < a < b$.

Note that in the above setting $P_{\tilde{\alpha}}^V$ is the distribution of a point process whose occurrence times T_i satisfy $T_i = T_{\tilde{\gamma}_i}^M$ for $i = 1, \ldots, n^*$ resp. $T_i = T_{\tilde{\gamma}_{n^*}+n^*-i}^M$ for $i > n^*$, where T_i^M are the occurrence times of a Poisson process mixed with V.

Alternative representations of I and w are given by

$$
I(\alpha_1, \ldots, \alpha_{n^*}) = \left\{ \tilde{\alpha} \in \mathbb{N}_0^{n^*} \;\middle|\; \sum_{j=i}^{n^*} \tilde{\alpha}_j \leq \sum_{j=i}^{n^*} \alpha_j, \, i = 2, \ldots, n^*; \; \sum_{j=1}^{n^*} \tilde{\alpha}_j = \sum_{j=1}^{n^*} \alpha_j \right\}
$$

and

$$
w\begin{pmatrix} \alpha_1, \ldots, \alpha_{n^*} \\ \tilde{\alpha}_1, \ldots, \tilde{\alpha}_{n^*} \end{pmatrix} = \frac{\prod_{i=1}^{n^*} \tilde{\alpha}_i! \gamma_i}{\Gamma(\gamma_{n^*}+1)} \prod_{i=2}^{n^*} \begin{pmatrix} \sum_{j=i}^{n^*} \alpha_i - \sum_{j=i+1}^{n^*} \tilde{\alpha}_i \\ \tilde{\alpha}_i \end{pmatrix}.
$$

Further, it is possible to include the case $n^* = 0$, i.e. constant parametrizing sequences (corresponding to $P_N = P^V$), in the formulation of the Proposition, interpreting appropriately the quantities $I = \emptyset$, $w() = 1$ and the sum.

Preceding the proof we study the following lemma which contains the main ideas of Theorem 46's proof:

Lemma 47: For $n \in \mathbb{N}$, $t_1, \ldots, t_n \in \mathbb{R}$, $t_0 = 0$ and $\alpha = (\alpha_1, \ldots, \alpha_n) \in \mathbb{N}_0^n$ we have

$$
\prod_{i=1}^n t_i^{\alpha_i} = \sum_{\tilde{\alpha} \in I(\alpha)} z\begin{pmatrix} \alpha \\ \tilde{\alpha} \end{pmatrix} \cdot \prod_{i=1}^n (t_i - t_{i-1})^{\tilde{\alpha}_i} \tag{4.9}
$$

where

$$
I(\alpha) = \left\{ \tilde{\alpha} \in \mathbb{N}_0^n \;\middle|\; \tilde{\gamma}_i \geq \gamma_i, \, i = 1, \ldots, n-1; \; \tilde{\gamma}_n = \gamma_n \right\}
$$

and

$$
z\begin{pmatrix} \alpha \\ \tilde{\alpha} \end{pmatrix} = \prod_{i=2}^n \begin{pmatrix} \tilde{\gamma}_i - \gamma_{i-1} - 1 \\ \tilde{\alpha}_i \end{pmatrix}.
$$

Proof: Fix $n \in \mathbb{N}$. Equation (4.9) is certainly true for $\alpha = (0, \ldots, 0)$ since

$$
I(0, \ldots, 0) = \{(0, \ldots, 0)\} \quad \text{and} \quad z\begin{pmatrix} 0, \ldots, 0 \\ 0, \ldots, 0 \end{pmatrix} = 1.
$$

Now, let equation (4.9) be true for some $(\alpha_1, \ldots, \alpha_n) \in \mathbb{N}_0^n$. Then for $(\alpha_1, \ldots, \alpha_{k-1}, \alpha_k + 1, \alpha_{k+1}, \ldots, \alpha_n)$ with $k \in \{1, \ldots, n\}$ we obtain

$$
\prod_{i=1}^n t_i^{\alpha_i} \cdot t_k = \prod_{i=1}^n t_i^{\alpha_i} \cdot \sum_{i=1}^k (t_i - t_{i-1}) = \sum_{\tilde{\alpha} \in I(\alpha)} z\begin{pmatrix} \alpha \\ \tilde{\alpha} \end{pmatrix} \cdot \prod_{i=1}^n (t_i - t_{i-1})^{\tilde{\alpha}_i} \cdot \sum_{i=1}^k (t_i - t_{i-1})
$$

for $t_1, \ldots, t_n \in \mathbb{R}$. Expanding the product of the two sums gives us again summands proportional to $\prod_{i=1}^n (t_i - t_{i-1})^{\tilde{\alpha}_i}$ with vectors $\tilde{\alpha}$ of exponents now lying within the set

$$
\left\{ \tilde{\alpha} \in \mathbb{N}_0^n \;\middle|\; \begin{array}{ll} \tilde{\gamma}_i \geq \gamma_i, & i = 1, \ldots, k-1, \\ \tilde{\gamma}_i \geq \gamma_i + 1, & i = k, \ldots, n-1, \\ \tilde{\gamma}_n = \gamma_n + 1, \end{array} \right\}
$$

which coincides with $I(\alpha_1, \ldots, \alpha_{k-1}, \alpha_k + 1, \alpha_{k+1}, \ldots, \alpha_n)$. Furthermore, for the multiplicities of each summand we subsequently prove that for $\tilde{\alpha} \in I(\alpha_1, \ldots, \alpha_{k-1}, \alpha_k + 1, \alpha_{k+1}, \ldots, \alpha_n)$ and $k = 1, \ldots, n$ we have

$$z\left(\begin{matrix} \alpha_1, \ldots, \alpha_{k-1}, \alpha_k + 1, \alpha_{k+1}, \ldots, \alpha_n \\ \tilde{\alpha}_1, \ldots, \tilde{\alpha}_{k-1}, \tilde{\alpha}_k, \tilde{\alpha}_{k+1}, \ldots, \tilde{\alpha}_n \end{matrix}\right) = \prod_{i=2}^{k} \binom{\tilde{\gamma}_i - \gamma_{i-1} - 1}{\tilde{\alpha}_i} \prod_{i=k+1}^{n} \binom{\tilde{\gamma}_i - \gamma_{i-1} - 2}{\tilde{\alpha}_i}.$$
(4.10)

Let initially $k = 1$ and $\tilde{\alpha} \in I(\alpha_1 + 1, \alpha_2, \ldots, \alpha_n)$. Then we find that $\tilde{\alpha}_1 \geq 1$ and that the summand $\prod_{i=1}^{n}(t_i - t_{i-1})^{\tilde{\alpha}_i}$ appears as often in the expansion of $t_1 \prod_{i=1}^{n} t_i^{\alpha_i}$ as the summand corresponding to $(\tilde{\alpha}_1 - 1, \tilde{\alpha}_2, \ldots, \tilde{\alpha}_n)$ appears in the expansion of $\prod_{i=1}^{n} t_i^{\alpha_i}$. Thus

$$z\left(\begin{matrix} \alpha_1 + 1, \alpha_2, \ldots, \alpha_n \\ \tilde{\alpha}_1, \tilde{\alpha}_2, \ldots, \tilde{\alpha}_n \end{matrix}\right) = z\left(\begin{matrix} \alpha_1, \alpha_2, \ldots, \alpha_n \\ \tilde{\alpha}_1 - 1, \tilde{\alpha}_2, \ldots, \tilde{\alpha}_n \end{matrix}\right) = \prod_{i=2}^{n} \binom{\tilde{\gamma}_i - \gamma_{i-1} - 2}{\tilde{\alpha}_i},$$

i.e., equation (4.10) holds for $k = 1$. Now assume equation (4.10) to hold for some $k \in \{1, \ldots, n-1\}$. For $k+1$ and $\tilde{\alpha} \in I(\alpha_1, \ldots, \alpha_k, \alpha_{k+1} + 1, \alpha_{k+2}, \ldots, \alpha_n)$ and due to

$$\prod_{i=1}^{n} t_i^{\alpha_i} \cdot t_{k+1} = \prod_{i=1}^{n} t_i^{\alpha_i} \cdot t_k + (t_{k+1} - t_k) \prod_{i=1}^{n} t_i^{\alpha_i}$$

we obtain as multiplicity of the summand corresponding to $\tilde{\alpha}$ if $\tilde{\alpha}_{k+1} = 0$

$$z\left(\begin{matrix} \alpha_1, \ldots, \alpha_k, \alpha_{k+1} + 1, \alpha_{k+2}, \ldots, \alpha_n \\ \tilde{\alpha}_1, \ldots, \tilde{\alpha}_k, 0, \tilde{\alpha}_{k+2}, \ldots, \tilde{\alpha}_n \end{matrix}\right) = z\left(\begin{matrix} \alpha_1, \ldots, \alpha_{k-1}, \alpha_k + 1, \alpha_{k+1}, \alpha_{k+2}, \ldots, \alpha_n \\ \tilde{\alpha}_1, \ldots, \tilde{\alpha}_{k-1}, \tilde{\alpha}_k, 0, \tilde{\alpha}_{k+2}, \ldots, \tilde{\alpha}_n \end{matrix}\right)$$

$$= \prod_{i=2}^{k} \binom{\tilde{\gamma}_i - \gamma_{i-1} - 1}{\tilde{\alpha}_i} \prod_{i=k+1}^{n} \binom{\tilde{\gamma}_i - \gamma_{i-1} - 2}{\tilde{\alpha}_i}$$

or if $\tilde{\alpha}_{k+1} > 0$

$$z\left(\begin{matrix} \alpha_1, \ldots, \alpha_k, \alpha_{k+1} + 1, \alpha_{k+2}, \ldots, \alpha_n \\ \tilde{\alpha}_1, \ldots, \tilde{\alpha}_k, \tilde{\alpha}_{k+1}, \tilde{\alpha}_{k+2}, \ldots, \tilde{\alpha}_n \end{matrix}\right)$$

$$= z\left(\begin{matrix} \alpha_1, \ldots, \alpha_{k-1}, \alpha_k + 1, \alpha_{k+1}, \ldots, \alpha_n \\ \tilde{\alpha}_1, \ldots, \tilde{\alpha}_{k-1}, \tilde{\alpha}_k, \tilde{\alpha}_{k+1}, \ldots, \tilde{\alpha}_n \end{matrix}\right) + z\left(\begin{matrix} \alpha_1, \ldots, \alpha_k, \alpha_{k+1}, \alpha_{k+2}, \ldots, \alpha_n \\ \tilde{\alpha}_1, \ldots, \tilde{\alpha}_k, \tilde{\alpha}_{k+1} - 1, \tilde{\alpha}_{k+2}, \ldots, \tilde{\alpha}_n \end{matrix}\right)$$

$$= \prod_{i=2}^{k} \binom{\tilde{\gamma}_i - \gamma_{i-1} - 1}{\tilde{\alpha}_i} \prod_{i=k+1}^{n} \binom{\tilde{\gamma}_i - \gamma_{i-1} - 2}{\tilde{\alpha}_i}$$

$$+ \prod_{i=2}^{k} \binom{\tilde{\gamma}_i - \gamma_{i-1} - 1}{\tilde{\alpha}_i} \binom{\tilde{\gamma}_{k+1} - \gamma_k - 2}{\tilde{\alpha}_{k+1} - 1} \prod_{i=k+2}^{n} \binom{\tilde{\gamma}_i - \gamma_{i-1} - 2}{\tilde{\alpha}_i}$$

$$= \prod_{i=2}^{k+1} \binom{\tilde{\gamma}_i - \gamma_{i-1} - 1}{\tilde{\alpha}_i} \prod_{i=k+2}^{n} \binom{\tilde{\gamma}_i - \gamma_{i-1} - 2}{\tilde{\alpha}_i},$$

since

$$\binom{\tilde{\gamma}_{k+1} - \gamma_k - 2}{\tilde{\alpha}_{k+1}} + \binom{\tilde{\gamma}_{k+1} - \gamma_k - 2}{\tilde{\alpha}_{k+1} - 1} = \binom{\tilde{\gamma}_{k+1} - \gamma_k - 1}{\tilde{\alpha}_{k+1}}$$

(the above equation is of the type $\binom{a}{b} + \binom{a}{b-1} = \binom{a+1}{b}$).

Hence, equation (4.10) also holds for $k+1$ and by induction is valid for $k = 1, \ldots, n$.
Altogether, equation (4.9) is true for arbitrary $n \in \mathbb{N}$ and $\alpha \in \mathbb{N}_0^n$ and the proof is
complete.   ∎

We can now proceed with the proof of Theorem 46:

Proof: We conduct the proof showing that the distribution $P_{T_1,\ldots,T_n}$ of $n \in \mathbb{N}$ successive occurrence times of N satisfies

$$P_{T_1,\ldots,T_n}(A) = \sum_{\tilde{\alpha} \in I} w\binom{\alpha_1, \ldots, \alpha_{n^*}}{\tilde{\alpha}_1, \ldots, \tilde{\alpha}_{n^*}} \cdot P_{T_1,\ldots,T_n}^{V,\tilde{\alpha}}(A), \qquad A \in \bar{\mathcal{B}}^n,$$

where $P_{T_1,\ldots,T_n}^{V,\tilde{\alpha}}$ is the joint distribution of $n \in \mathbb{N}$ successive occurrence times of a
point process following the distribution $P_{\tilde{\alpha}}^V$. For $n \geq n^*$, we have according to
Theorem 40

$$P_{T_1,\ldots,T_n}^{W}(A) = P(T_1 < \infty) \int_A f_{T_1,\ldots,T_n} d\bar{\ell}^n + P(T_1 = \infty)\delta_{(\infty,\ldots,\infty)}(A), \qquad A \in \bar{\mathcal{B}}^n,$$

where

$$f_{T_1,\ldots,T_n}(t_1,\ldots,t_n) = \frac{\prod_{i=1}^n \gamma_i}{\Gamma(\gamma_n + 1)} \prod_{i=1}^n t_i^{\alpha_i} \int_{(0,\infty)} \lambda^{\gamma_n} e^{-\lambda t_n}\, dW(\lambda), \quad 0 < t_1 \leq \cdots \leq t_n.$$

Lemma 47 yields

$$f_{T_1,\ldots,T_n}(t_1,\ldots,t_n)$$

$$= \sum_{\tilde{\alpha} \in I(\alpha_1,\ldots,\alpha_n)} \frac{\prod_{i=1}^n \gamma_i}{\Gamma(\gamma_n + 1)} \cdot z\binom{\alpha}{\tilde{\alpha}} \cdot \prod_{i=1}^n (t_i - t_{i-1})^{\tilde{\alpha}_i} \int_{(0,\infty)} \lambda^{\gamma_n} e^{-\lambda t_n}\, dW(\lambda)$$

$$= \sum_{\tilde{\alpha} \in I(\alpha_1,\ldots,\alpha_n)} \frac{\prod_{i=1}^n \gamma_i}{\Gamma(\gamma_n + 1)} \cdot z\binom{\alpha}{\tilde{\alpha}} \prod_{i=1}^n \tilde{\alpha}_i! \cdot \underbrace{\frac{\prod_{i=1}^n (t_i - t_{i-1})^{\tilde{\alpha}_i}}{\prod_{i=1}^n \tilde{\alpha}_i!} \int_{(0,\infty)} \lambda^{\gamma_n} e^{-\lambda t_n}\, dW(\lambda)}_{=f_n^{[n]}(t_1,\ldots,t_n|\tilde{\alpha}_1,\ldots,\tilde{\alpha}_n)}$$

$$(4.11)$$

for $n \geq n^*$ and $0 = t_0 < t_1 \leq \cdots \leq t_n$. Thus defined, $f_n^{[n]}(t_1,\ldots,t_n|\tilde{\alpha}_1,\ldots,\tilde{\alpha}_n)$ is a
density concentrated on $\{(t_1,\ldots,t_n) \in \mathbb{R}^n | 0 \leq t_1 < \cdots < t_n\}$. Even more, according
to Lemma 12 it is the joint density of occurrence times of a mixed Poisson process

with mixing distribution W, provided that at least one jump occurs, after deleting the points $T_{\tilde{\gamma}_{i-1}+1}, \ldots, T_{\tilde{\gamma}_i-1}$ for $\tilde{\gamma}_{i-1} + 1 \leq \tilde{\gamma}_i - 1$, $i = 1, \ldots, n$.

Since $\alpha_i = 0$ and thus $\gamma_{i-1} + 1 > \gamma_i - 1$ for $i > n^*$ we actually delete only the points $T_{\tilde{\gamma}_{i-1}+1}, \ldots, T_{\tilde{\gamma}_i-1}$ for $i = 1, \ldots, n^*$. Additionally, $\alpha_i = 0$ for $i > n^*$ implies for $n \geq n^*$

$$f_n^{[n]}(t_1, \ldots, t_n | \tilde{\alpha}_1, \ldots, \tilde{\alpha}_n) \;=\; f_n^{[n^*]}(t_1, \ldots, t_n | \tilde{\alpha}_1, \ldots, \tilde{\alpha}_{n^*}), \qquad t_1, \ldots, t_n \in \bar{\mathbb{R}},$$

further

$$(\tilde{\alpha}_1, \ldots, \tilde{\alpha}_n) \in I(\alpha_1, \ldots, \alpha_n) \Leftrightarrow (\tilde{\alpha}_1, \ldots, \tilde{\alpha}_{n^*}) \in I(\alpha_1, \ldots, \alpha_{n^*}) \text{ and } \tilde{\alpha}_i = 0, \, i > n^*,$$

and

$$z\begin{pmatrix} \alpha_1, \ldots, \alpha_n \\ \tilde{\alpha}_1, \ldots, \tilde{\alpha}_n \end{pmatrix} \;=\; z\begin{pmatrix} \alpha_1, \ldots, \alpha_{n^*} \\ \tilde{\alpha}_1, \ldots, \tilde{\alpha}_{n^*} \end{pmatrix} \quad \text{and} \quad \frac{\prod_{i=1}^n \gamma_i}{\Gamma(\gamma_n + 1)} \;=\; \frac{\prod_{i=1}^{n^*} \gamma_i}{\Gamma(\gamma_{n^*} + 1)}.$$

As

$$\frac{\prod_{i=1}^{n^*} \tilde{\alpha}_i! \gamma_i}{\Gamma(\gamma_{n^*} + 1)} \cdot z\begin{pmatrix} \alpha_1, \ldots, \alpha_{n^*} \\ \tilde{\alpha}_1, \ldots, \tilde{\alpha}_{n^*} \end{pmatrix} \;=\; \frac{\prod_{i=1}^{n^*} \tilde{\alpha}_i! \gamma_i}{\Gamma(\gamma_{n^*} + 1)} \prod_{i=2}^{n^*} \begin{pmatrix} \tilde{\gamma}_i - \gamma_{i-1} - 1 \\ \tilde{\alpha}_i \end{pmatrix}$$

$$= \; w\begin{pmatrix} \alpha_1, \ldots, \alpha_{n^*} \\ \tilde{\alpha}_1, \ldots, \tilde{\alpha}_{n^*} \end{pmatrix}$$

holds for the weights w which sum up to 1 (this follows from integrating equation (4.11) for $n = n^*$), we altogether find

$$P_{T_1,\ldots,T_n}(A)$$

$$= \sum_{\tilde{\alpha} \in I(\alpha_1,\ldots,\alpha_{n^*})} w\begin{pmatrix} \alpha_1, \ldots, \alpha_{n^*} \\ \tilde{\alpha}_1, \ldots, \tilde{\alpha}_{n^*} \end{pmatrix} \cdot$$

$$\left(P(T_1 < \infty) \int_A f_n^{[n^*]}(t_1, \ldots, t_n | \tilde{\alpha}_1, \ldots, \tilde{\alpha}_{n^*}) d\bar{\ell}^n + P(T_1 = \infty)\delta_{(\infty,\ldots,\infty)}(A) \right)$$

$$= \sum_{\tilde{\alpha} \in I(\alpha_1,\ldots,\alpha_{n^*})} w\begin{pmatrix} \alpha_1, \ldots, \alpha_{n^*} \\ \tilde{\alpha}_1, \ldots, \tilde{\alpha}_{n^*} \end{pmatrix} P_{T_1,\ldots,T_n}^{V,\tilde{\alpha}}(A), \qquad A \in \bar{\mathcal{B}}^n, \, n \geq n^*,$$

where the last equality holds due to Lemma 12. Since the distribution P_N of a point process N is completely described by the distributions $P_{T_1,\ldots,T_n}$ for $n \geq n^*$ the proof is complete. ∎

Example 48: In the situation of Theorem 46, let us consider especially $n^* = 2$. Then we find

$$I \;=\; \left\{ (\tilde{\alpha}_1, \tilde{\alpha}_2) \in \mathbb{N}_0^2 \,|\, \tilde{\alpha}_2 \leq \alpha_2 \,;\, \tilde{\alpha}_1 + \tilde{\alpha}_2 = \alpha_1 + \alpha_2 \right\}$$

$$= \left\{ (\alpha_1 + \alpha_2 - i, i) | i = 0, \ldots, \alpha_2 \right\},$$

$$w\begin{pmatrix} \alpha_1, \alpha_2 \\ \alpha_1 + \alpha_2 - i, i \end{pmatrix} = \frac{(\alpha_1 + \alpha_2 - i)!i!(\alpha_1 + 1)(\alpha_1 + \alpha_2 + 2)}{(\alpha_1 + \alpha_2 + 2)!}\begin{pmatrix} \alpha_2 \\ i \end{pmatrix}$$

$$= \frac{(\alpha_1 + 1)!\alpha_2!}{(\alpha_1 + \alpha_2 + 1)!} \cdot \frac{(\alpha_1 + \alpha_2 - i)!}{(\alpha_2 - i)!\alpha_1!} = \frac{\binom{\alpha_1 + \alpha_2 - i}{\alpha_1}}{\binom{\alpha_1 + \alpha_2 + 1}{\alpha_2}}$$

and

$$P_N(A) = \sum_{i=0}^{\alpha_2} \frac{\binom{\alpha_1 + \alpha_2 - i}{\alpha_1}}{\binom{\alpha_1 + \alpha_2 + 1}{\alpha_2}} P_{\alpha_1 + \alpha_2 - i, i}^V(A), \qquad A \in \mathcal{B}(\mathcal{N}). \qquad \square$$

Let us apply Theorem 46 in order to find a decomposition of a GOS-process in terms of modified mixed Poisson processes for an example with $n^* = 3$:

Example 49: Let $\alpha_1 = \alpha_3 = 1$ and $\alpha_i = 0$ else. Let N be a GOS-process w.r.t. the given sequence $\{\alpha_i\}_{i \in \mathbb{N}}$. Hence, $n^* = 3$ and we obtain

$$I(1,0,1) = \{(1,0,1),(1,1,0),(2,0,0)\}.$$

For the weights w of Theorem 46 we obtain

$$w\begin{pmatrix} 1 & 0 & 1 \\ 1 & 0 & 1 \end{pmatrix} \overset{(4.8)}{=} \frac{1! \, 0! \, 1! \cdot 2 \cdot 3 \cdot 5}{5!} \cdot \begin{pmatrix} 3-2-1 \\ 0 \end{pmatrix} \cdot \begin{pmatrix} 5-3-1 \\ 1 \end{pmatrix} = \frac{1}{4}$$

and analogously

$$w\begin{pmatrix} 1 & 0 & 1 \\ 1 & 1 & 0 \end{pmatrix} \overset{(4.8)}{=} \frac{1}{4} \quad \text{and} \quad w\begin{pmatrix} 1 & 0 & 1 \\ 2 & 0 & 0 \end{pmatrix} \overset{(4.8)}{=} \frac{1}{2}.$$

The summand indexed by $\tilde{\alpha} = (1,0,1)$ with $\tilde{\gamma}_0 = 0$, $\tilde{\gamma}_1 = 2$, $\tilde{\gamma}_2 = 3$ and $\tilde{\gamma}_3 = 5$ corresponds to a Poisson process after the deletion of its points $T_{\tilde{\gamma}_0+1}, \ldots, T_{\tilde{\gamma}_1-1}$, $T_{\tilde{\gamma}_1+1}, \ldots, T_{\tilde{\gamma}_2-1}$ and $T_{\tilde{\gamma}_2+1}, \ldots, T_{\tilde{\gamma}_3-1}$, that is after deletion of T_1 and T_4. Altogether, we can represent N as a mixture of

- a mixed Poisson process whose first and third point are deleted, with a probability of $\frac{1}{4}$,

- a mixed Poisson process whose first and fourth point are deleted, with a probability of $\frac{1}{4}$, and

- a mixed Poisson process whose first and second point are deleted, with a probability of $\frac{1}{2}$. $\qquad \square$

Via time transformations it is possible to extend Theorem 46 to certain GOS-processes whose parametrizing sequence is eventually constant but nonzero:

Corollary 50: *Let $\{\beta_i\}_{i \in \mathbb{N}} \subset \mathbb{N}_0$ be a sequence of natural numbers such that $\beta_i = 0$ for $i > n^*$ and some $n^* \in \mathbb{N}_0$. Put $\alpha_i = (\beta_i + 1) \cdot (\alpha + 1) - 1$ for $i \in \mathbb{N}$ and some $\alpha > -1$. Further let N be a generalized order statistic process with respect to $\{\alpha_i\}_{i \in \mathbb{N}}$, i. e., its successive occurrence times $T_1, T_2, \ldots$, given $T_1 < \infty$, follow the joint densities*

$$f_{T_1,\ldots,T_n}(t_1,..,t_n) = \frac{\prod_{i=1}^{n} \gamma_i t_i^{\alpha_i}}{\Gamma\left(\frac{\gamma_n}{\alpha+1} + 1\right)} \int_{(0,\infty)} \lambda^{\frac{\gamma_n}{\alpha+1}} e^{-\lambda t_n^{\alpha+1}} \, dW(\lambda), \quad n \geq n^*, 0 < t_1 \leq \cdots \leq t_n,$$

for some probability distribution W on $(0, \infty)$.

Then, the distribution P_N of N verifies

$$P_N(A) = \sum_{\tilde{\alpha} \in I} w\binom{\beta_1, \ldots, \beta_{n^*}}{\tilde{\alpha}_1, \ldots, \tilde{\alpha}_{n^*}} \cdot P_{\tilde{\alpha}}^{V,\alpha}(A), \qquad A \in \mathcal{H}(\mathcal{N}),$$

with index set

$$I = I(\beta_1, \ldots, \beta_{n^*}) = \left\{ \tilde{\alpha} \in \mathbb{N}_0^{n^*} \,\middle|\, \tilde{\gamma}_i \geq \delta_i, i = 1, \ldots, n^* - 1 \,;\, \tilde{\gamma}_{n^*} = \delta_{n^*} \right\}$$

and weights

$$w\binom{\beta_1, \ldots, \beta_{n^*}}{\tilde{\alpha}_1, \ldots, \tilde{\alpha}_{n^*}} = \frac{\prod_{i=1}^{n^*} \tilde{\alpha}_i! \delta_i}{\Gamma(\delta_{n^*} + 1)} \prod_{i=2}^{n^*} \binom{\tilde{\gamma}_i - \delta_{i-1} - 1}{\tilde{\alpha}_i}, \qquad (4.12)$$

where $\delta_i = \sum_{j=1}^{i}(\beta_j + 1)$, $i \in \mathbb{N}$, and where $P_{\tilde{\alpha}}^{V,\alpha}$ denotes the distribution of a mixed Poisson process with mixing distribution V undergoing the time transformation $t \mapsto t^{\alpha+1}$ and deleting the points $T_{\tilde{\gamma}_{i-1}+1}, \ldots, T_{\tilde{\gamma}_i - 1}$ for $\tilde{\gamma}_{i-1} + 1 \leq \tilde{\gamma}_i - 1$, $i = 1, \ldots, n^*$. The distribution V is as in Theorem 46.

Proof: Let M be the point process given by $M(t) = N\left(t^{\frac{1}{\alpha+1}}\right)$, $t \geq 0$. According to Proposition 34, M is a GOS-process w.r.t. $\{\beta_i\}_{i \in \mathbb{N}}$ – note that $\beta_i = \frac{((\beta_i+1)(\alpha+1)-1)+1}{\alpha+1} - 1$ for $i = 1, \ldots, n^*$ – and with corresponding densities

$$f_{T_1,\ldots,T_n}(t_1, \ldots, t_n) = \frac{\prod_{i=1}^{n} \delta_i t_i^{\beta_i}}{\Gamma(\delta_n + 1)} \int_{(0,\infty)} \lambda^{\delta_n} e^{-\lambda t_n} \, dW(\lambda)$$

for $n \geq n^*$, $0 < t_1 \leq \cdots \leq t_n$. Theorem 46 applied to M and a back transformation of time yield the result. ∎

Remark that though the studied GOS-processes can be obtained from mixed Poisson processes deleting some points, in general a mixed Poisson process with some absent points does not satisfy the generalized order statistic property:

Example 51: Consider a point process with independent sojourn times S_i, $i \in \mathbb{N}$, such that S_1 and S_2 are $\Gamma(\lambda, 2)$-distributed. This holds especially for Poisson processes with intensity λ whose first and third point are deleted. Then we find

$$
\begin{aligned}
f_{T_1, T_2}(t_1, t_2) &= f_{S_1}(t_1) f_{S_2}(t_2 - t_1) \\
&= \lambda^2 t_1 e^{-\lambda t_1} \cdot \lambda^2 (t_2 - t_1) e^{-\lambda(t_2 - t_1)} \\
&= \lambda^4 t_1 (t_2 - t_1) e^{-\lambda t_2}, \qquad 0 < t_1 \le t_2.
\end{aligned}
$$

Due to Theorem 32 the process cannot satisfy the generalized order statistic property since the density does not admit a structure of the form $t_1^{\alpha_1} t_2^{\alpha_2} \varphi_2(t_2)$. $\qquad\square$

An asymptotic result

Since for mixed Poisson processes it is well known that the ratio $\frac{n}{T_n}$ converges to a random variable which follows the mixing distribution of the process, there is reasonable hope to be able to establish a similar result in the context of generalized order statistic processes. In the sequel we exploit the decomposition of GOS-processes obtained in the previous section to deduce a proposition concerning the limiting distribution of the ratio $\frac{\delta_i}{T_i^{\alpha+1}}$ of a GOS-process whose parametrizing sequence is eventually constant $\alpha > -1$:

Corollary 52: *With the notations and under the assumptions of Corollary 50 we find*

$$
\lim_{i \to \infty} P\left(\frac{\delta_i}{T_i^{\alpha+1}} \le x \right) = V(x), \qquad x \in \mathbb{R},
$$

if x is a continuity point of V.

Note that we interpret the mixing distribution V simultaneously as distribution function.

Proof: The proof will be traced back to the fact that for mixed Poisson processes with mixing distribution V the distribution of the ratio of i and its i-th occurrence time converges to V.

Let $P_{\tilde{\alpha}}^{V, \alpha}$ as in Corollary 50 denote the distribution of a mixed Poisson process with mixing distribution V, undergoing the time transformation $t \mapsto t^{\alpha+1}$ and deleting the $\tilde{\gamma}_{n^*} - n^*$ process points $T_{\tilde{\gamma}_{i-1}+1}, \ldots, T_{\tilde{\gamma}_i - 1}$ for $\tilde{\gamma}_{i-1} + 1 \le \tilde{\gamma}_i - 1$, $i = 1, \ldots, n^*$. Further, let $P^{V, \alpha} = P_0^{V, \alpha}$ be the distribution of a mixed Poisson process with mixing distribution V and time transformation $t \mapsto t^{\alpha+1}$. In particular, $P^V = P^{V, 0}$ is the distribution of a mixed Poisson process with mixing distribution V whose occurrence times we denote by T_i^V. For a realization $\mathfrak{n} \in \mathcal{N}$ of a point process we denote by $\mathfrak{t}_i = \mathfrak{t}_i(\mathfrak{n})$, $i \in \mathbb{N}$, the corresponding realizations of occurrence times.

Then we find:

$$\lim_{i\to\infty} P\left(\frac{\delta_i}{T_i^{\alpha+1}} \leq x\right)$$

$$= \lim_{i\to\infty} P_N\left(\left\{\mathfrak{n}\in\mathcal{N}\,\middle|\,\frac{\delta_i}{\mathfrak{t}_i(\mathfrak{n})^{\alpha+1}} \leq x\right\}\right)$$

$$= \lim_{i\to\infty} \sum_{\tilde{\alpha}\in I} w\binom{\beta_1,\ldots,\beta_{n^*}}{\tilde{\alpha}_1,\ldots,\tilde{\alpha}_{n^*}} \cdot P_{\tilde{\alpha}}^{V,\alpha}\left(\left\{\mathfrak{n}\in\mathcal{N}\,\middle|\,\frac{\delta_i}{\mathfrak{t}_i(\mathfrak{n})^{\alpha+1}} \leq x\right\}\right)$$

$$= \lim_{i\to\infty} \sum_{\tilde{\alpha}\in I} w\binom{\beta_1,\ldots,\beta_{n^*}}{\tilde{\alpha}_1,\ldots,\tilde{\alpha}_{n^*}} \cdot P^{V,\alpha}\left(\left\{\mathfrak{n}\in\mathcal{N}\,\middle|\,\frac{\delta_i}{\mathfrak{t}_{\tilde{\gamma}_i}(\mathfrak{n})^{\alpha+1}} \leq x\right\}\right)$$

$$= \lim_{i\to\infty} \sum_{\tilde{\alpha}\in I} w\binom{\beta_1,\ldots,\beta_{n^*}}{\tilde{\alpha}_1,\ldots,\tilde{\alpha}_{n^*}} \cdot P^V\left(\left\{\mathfrak{n}\in\mathcal{N}\,\middle|\,\frac{\delta_i}{\mathfrak{t}_{\tilde{\gamma}_i}(\mathfrak{n})} \leq x\right\}\right)$$

$$= \sum_{\tilde{\alpha}\in I} w\binom{\beta_1,\ldots,\beta_{n^*}}{\tilde{\alpha}_1,\ldots,\tilde{\alpha}_{n^*}} \cdot \lim_{i\to\infty} P\left(\frac{\delta_i}{T_{\tilde{\gamma}_i}^V} \leq x\right)$$

$$= \sum_{\tilde{\alpha}\in I} w\binom{\beta_1,\ldots,\beta_{n^*}}{\tilde{\alpha}_1,\ldots,\tilde{\alpha}_{n^*}} \cdot V(x) = V(x),$$

since $\tilde{\gamma}_i = \tilde{\gamma}_{n^*} + i - n^* = \delta_{n^*} + i - n^* = \delta_i$ for $i \geq n^*$ and $\delta_i \to \infty$ for $i \to \infty$. ∎

Corollary 52 reflects that the path of a GOS-process w.r.t. an eventually constant sequence behaves finally like a time transformed mixed Poisson process – a fact which we already exploited in Section 4.2, especially Proposition 43.

4.5 Different representations of occurrence time densities

In view of the results gained until now for generalized order statistic processes with respect to eventually constant sequences, the idea comes up to achieve some generalizations that also hold for arbitrary parametrizing sequences $\{\alpha_i\}_{i\in\mathbb{N}}$ by letting n^* tend to infinity, where n^* denotes the index from which on the elements of $\{\alpha_i\}_{i\in\mathbb{N}}$ are constant, thus connecting arbitrary parametrizing sequences with those eventually constant. In particular, we would like to deduce a representation analogous to the one given for eventually constant parametrizing sequences in Theorem 40. The actual section provides preliminary results necessary in order to do so, but does not itself present any generalizations, for what we refer to Chapter 5 in which the above idea is exploited in order to explicitly specify the distribution of a GOS-process in the case of increasing convergent parametrizing sequences, essentially.

Our actual studies originate in the representation of the distribution of a GOS-process w. r. t. $\{\alpha_i\}_{i \in \mathbb{N}}$ such that $\alpha_i = \alpha$ for $i > n^* \in \mathbb{N}$. Theorem 40 states that its distribution is given by the probability $P(T_1 < \infty)$ and the following n-dimensional densities of its occurrence times $T_1, \ldots, T_n$ for $n \in \mathbb{N}$, provided $T_1 < \infty$:

$$f_{T_1, \ldots, T_n}(t_1, \ldots, t_n) \;=\; \prod_{i=1}^{n} t_i^{\alpha_i} \varphi_n(t_n), \qquad 0 < t_1 \leq \cdots \leq t_n,$$

where

$$\varphi_n(t) \tag{4.13}$$

$$= \begin{cases} \dfrac{\lambda^{\frac{\gamma_n}{\alpha+1}} \prod_{i=1}^{n} \gamma_i}{\Gamma\left(\frac{\gamma_n}{\alpha+1}+1\right)} e^{-\lambda t^{\alpha+1}} & \text{for } n \geq n^*, \\[3ex] \dfrac{\lambda^{\frac{\gamma_{n^*}}{\alpha+1}} \prod_{i=1}^{n^*} \gamma_i}{\Gamma\left(\frac{\gamma_{n^*}}{\alpha+1}+1\right)} \displaystyle\int_t^\infty \int_{s_{n+1}}^\infty \cdots \int_{s_{n^*-1}}^\infty \prod_{i=n+1}^{n^*} s_i^{\alpha_i} \cdot e^{-\lambda s_{n^*}^{\alpha+1}} \, ds_{n^*} \cdots ds_{n+1} & \text{else} \end{cases}$$

for $t > 0$. In the above representation of φ_n and for what remains of the section we neglect the presence of a mixing distribution W since the central point is not the mixture but the analysis and improved understanding of the basic distributional model. However, in what follows, it would in principle be possible to carry along the mixed structure.

Thinking of n^* tending to infinity, the iterated integrals in (4.13) are quite inconvenient. Thus it is our objective to find alternative representations of φ_n that later will turn out to be more suitable concerning a limit approach.

Put

$$\Phi_\lambda^n(t \mid a_1, \ldots, a_n; a) \;=\; \int_t^\infty \int_{s_1}^\infty \cdots \int_{s_{n-1}}^\infty \prod_{i=1}^{n} s_i^{a_i} \cdot e^{-\lambda s_n^{a+1}} \, ds_n \cdots ds_1, \qquad n \in \mathbb{N},$$

$$\tag{4.14}$$

and

$$\Phi_\lambda^0(t \mid a) \;=\; e^{-\lambda t^{a+1}}$$

for $t > 0$, $\lambda > 0$ and $a > -1$. With the help of Φ_λ^n we may write

$$f_{T_1, \ldots, T_n}(t_1, \ldots, t_n) \;=\; \frac{\lambda^{\frac{\gamma_{n^*}}{\alpha+1}} \prod_{i=1}^{n^*} \gamma_i}{\Gamma\left(\frac{\gamma_{n^*}}{\alpha+1} + 1\right)} \prod_{i=1}^{n} t_i^{\alpha_i} \cdot \Phi_\lambda^{n^*-n}(t_n \mid \alpha_{n+1}, \ldots, \alpha_{n^*}; \alpha),$$

where $n \in \mathbb{N}$ such that $n \leq n^*$ and $0 < t_1 \leq \cdots \leq t_n$. Apparently, (4.14) yields the following recursion formula corresponding to (3.10) for $n \in \mathbb{N}$, $t > 0$:

$$\Phi_\lambda^n(t \mid a_1, \ldots, a_n; a) \;=\; \int_t^\infty s^{a_1} \Phi_\lambda^{n-1}(s \mid a_2, \ldots, a_n; a) \, ds. \tag{4.15}$$

In the sequel, we will need the following quantities: Given a sequence $\{a_i\}_{i \in \mathbb{N}} \subset \mathbb{R}$ and a real number $a > -1$ we put

$$A_j^k \;=\; \begin{cases} \sum_{i=j}^{k} \frac{a_i+1}{a+1} & \text{if } j \leq k, \\[1.5ex] 0 & \text{else.} \end{cases}$$

The following two propositions state alternative representations of Φ_λ^n:

Proposition 53: *Let $\{a_i\}_{i\in\mathbb{N}} \in \mathbb{R}$ be a sequence and $a > -1$. Then for $\lambda > 0$, $t > 0$ and $n \in \mathbb{N}_0$*

$$\Phi_\lambda^n\left(\left(\frac{t}{\lambda}\right)^{\frac{1}{a+1}}\bigg|\, a_1,\dots,a_n; a\right) = \left(\frac{t}{\lambda}\right)^{A_1^n}(a+1)^{-n}\cdot\frac{1}{2\pi\mathrm{i}}\int_{c-\mathrm{i}\infty}^{c+\mathrm{i}\infty} t^{-z}\frac{\Gamma(z)}{\prod_{i=1}^n\left(z - A_i^n\right)}\,dz,$$
$$(4.16)$$

where $c > 0$ and $c > A_i^n$ for $i = 1,\dots,n$.

The integral on the right side is a complex contour integral whose path of integration is a vertical line crossing the real axis in c.

Proof: Denote by $\Re(z)$ resp. $\Im(z)$ the real resp. imaginary part of z. We apply Mellin transform calculus, see Appendix B.2, and show by induction on $n \in \mathbb{N}_0$ that $\hat{f}^M(z) = \frac{\Gamma(z)}{\prod_{i=1}^n\left(z-A_i^n\right)}$ is the Mellin transform of the function

$$f(t) = \left(\frac{t}{\lambda}\right)^{-A_1^n}(a+1)^n\,\Phi_\lambda^n\left(\left(\frac{t}{\lambda}\right)^{\frac{1}{a+1}}\bigg|\, a_1,\dots,a_n; a\right).$$

Then the statement holds applying the inversion formula for Mellin transforms, compare equation (B.2).

For $n = 0$ equation (4.16) becomes

$$e^{-t} = \Phi_\lambda^0\left(\left(\frac{t}{\lambda}\right)^{\frac{1}{a+1}}\bigg|\, a\right) = \frac{1}{2\pi\mathrm{i}}\int_{c-\mathrm{i}\infty}^{c+\mathrm{i}\infty} t^{-z}\Gamma(z)\,dz, \qquad t > 0,$$

which is true since $\Gamma(z)$ with $\Re(z) > 0$ is the Mellin transform of e^{-t}. Now assume equation (4.16) to be valid for some $n \in \mathbb{N}_0$. Note, that $(z+b)^{-1}$ with $\Re(z) > -\Re(b)$ is the Mellin transform of $t^b \mathbb{1}_{(0,1)}(t)$ for $b \in \mathbb{C}$, where $\mathbb{1}_{(0,1)}$ denotes the indicator function of the interval $(0,1)$. Then we find for $t > 0$

$$\frac{1}{2\pi\mathrm{i}}\int_{c-\mathrm{i}\infty}^{c+\mathrm{i}\infty} t^{-z}\frac{\Gamma(z)}{\prod_{i=1}^{n+1}\left(z - A_i^{n+1}\right)}\,dz$$

$$= \frac{1}{2\pi\mathrm{i}}\int_{c-\mathrm{i}\infty}^{c+\mathrm{i}\infty} t^{-z}\frac{\Gamma(z)}{\prod_{i=2}^{n+1}\left(z - A_i^{n+1}\right)}\cdot\frac{1}{z - A_1^{n+1}}\,dz$$

$$\stackrel{(B.3)}{=} \int_0^\infty \left(\frac{t}{s}\right)^{-A_1^{n+1}}\mathbb{1}_{(0,1)}\left(\frac{t}{s}\right)\left(\frac{s}{\lambda}\right)^{-A_2^{n+1}}(a+1)^n\Phi_\lambda^n\left(\left(\frac{s}{\lambda}\right)^{\frac{1}{a+1}}\bigg|\, a_2,\dots,a_{n+1}; a\right)\frac{ds}{s}$$

$$= t^{-A_1^{n+1}}\lambda^{A_2^{n+1}}(a+1)^n\int_t^\infty s^{A_1^1 - 1}\,\Phi_\lambda^n\left(\left(\frac{s}{\lambda}\right)^{\frac{1}{a+1}}\bigg|\, a_2,\dots,a_{n+1}; a\right)ds = (*).$$

We substitute $s = \lambda \tilde{s}^{a+1}$ with $ds = \lambda(a+1)\tilde{s}^a \, d\tilde{s}$ which yields

$$
(*) \quad = \quad t^{-A_1^{n+1}} \lambda^{A_2^{n+1}} (a+1)^n \int_{\left(\frac{t}{\lambda}\right)^{\frac{1}{a+1}}}^{\infty} \left(\lambda \tilde{s}^{a+1}\right)^{A_1^1 - 1} \Phi_\lambda^n\left(\tilde{s} \,\Big|\, a_2, \ldots, a_{n+1}; a\right) \lambda(a+1)\tilde{s}^a \, d\tilde{s}
$$

$$
= \quad t^{-A_1^{n+1}} \lambda^{A_1^{n+1}} (a+1)^{n+1} \int_{\left(\frac{t}{\lambda}\right)^{\frac{1}{a+1}}}^{\infty} \tilde{s}^{a_1} \Phi_\lambda^n\left(\tilde{s} \,\Big|\, a_2, \ldots, a_{n+1}; a\right) d\tilde{s}
$$

$$
\overset{(4.15)}{=} \quad t^{-A_1^{n+1}} \lambda^{A_1^{n+1}} (a+1)^{n+1} \Phi_\lambda^{n+1}\left(\left(\frac{t}{\lambda}\right)^{\frac{1}{a+1}} \,\Big|\, a_1, a_2, \ldots, a_{n+1}; a\right), \qquad t > 0.
$$

That is, equation (4.16) holds for $n+1$ and the proposition is proven. ∎

We have represented Φ_λ^n as a so called Mellin-Barnes integral. Similar formulas of some familiar special functions can be found in the book of Paris and Kaminski (2001). Further, notice that Φ_λ^n is connected to Meijer's G-function, see Appendix A.3. More precisely, we find by definition of Meijer's G-function

$$
\left(\frac{t}{\lambda}\right)^{-A_1^n} (a+1)^n \, \Phi_\lambda^n\left(\left(\frac{t}{\lambda}\right)^{\frac{1}{a+1}} \,\Big|\, a_1, \ldots, a_n; a\right)
$$

$$
= \quad G_{n,n+1}^{n+1,0}\left(t \,\Big|\, \begin{matrix} 1 - A_1^n, 1 - A_2^n, \ldots, 1 - A_n^n \\ 0, -A_1^n, \ldots, -A_n^n \end{matrix}\right), \qquad t > 0,
$$

where the contour needed for the integral defining the above Meijer's G-function can in this case be chosen to be a vertical line crossing the real axis in a point exceeding 0 and A_j^n, $j = 1, \ldots, n$, compare (Mathai, 1993, p. 62).

The next proposition provides a series representation:

Proposition 54: Let $\{a_i\}_{i \in \mathbb{N}} \subset \mathbb{R}$ be such that $a_i = a$, $i > n$, for some $n \in \mathbb{N}_0$ and $a > -1$. If further $A_1^n, \ldots, A_n^n$ are mutually distinct, none of it being 0 or a negative integer, then we have for $t > 0$

$$
t^{A_1^n} \cdot \frac{1}{2\pi i} \int_{c-i\infty}^{c+i\infty} t^{-z} \frac{\Gamma(z)}{\prod_{i=1}^n (z - A_i^n)} \, dz = \sum_{k=0}^{\infty} (-1)^k \frac{t^{A_1^k} \Gamma\left(A_{k+1}^n + 1\right)}{\prod_{i=1}^k A_i^k \prod_{i=k+1}^n A_{k+1}^i} \tag{4.17}
$$

where $c > A_1^n$.

Remark the similarities between the above series and the exponential series $e^{-t} = \sum_{k=0}^{\infty} (-1)^k \frac{t^k}{k!}$ where A_1^k corresponds to k and likewise $\prod_{i=1}^k A_i^k$ corresponds to $k!$. This interpretation is reasonable because for large k we have $A_i^k - A_i^{k-1} = \frac{a_{k+1}}{a+1} = 1$ and as we will see later $\frac{\Gamma\left(A_{k+1}^n + 1\right)}{\prod_{i=k+1}^n A_{k+1}^i}$ is bounded, compare Lemma 64.

Proof: Let $t > 0$, $n \in \mathbb{N}_0$ and $a_1, \ldots, a_n, a > -1$ throughout the proof. The right hand side of equation (4.17) equals

$$\sum_{k=0}^{n-1} (-1)^k \frac{t^{A_1^k} \Gamma\left(A_{k+1}^n + 1\right)}{\prod_{i=1}^k A_i^k \prod_{i=k+1}^n A_{k+1}^i} + \sum_{k=n}^{\infty} (-1)^k \frac{t^{A_1^k} \Gamma\left(A_{k+1}^n + 1\right)}{\prod_{i=1}^k A_i^k \prod_{i=k+1}^n A_{k+1}^i}. \tag{4.18}$$

Taking into account that $A_{k+1}^n = 0$ and $A_i^k = A_i^n + k - n$ hold for $i \leq n \leq k$ and that $A_i^k = k - i + 1$ for $n < i \leq k$ we obtain for the right sum

$$\sum_{k=n}^{\infty} (-1)^k \frac{t^{A_1^k} \Gamma\left(A_{k+1}^n + 1\right)}{\prod_{i=1}^k A_i^k \prod_{i=k+1}^n A_{k+1}^i} = (-1)^n \sum_{k=n}^{\infty} (-1)^{k-n} \frac{t^{A_1^n + k - n}}{\prod_{i=1}^n A_i^k \prod_{i=n+1}^k A_i^k}$$

$$= (-1)^n t^{A_1^n} \sum_{k=n}^{\infty} \frac{(-t)^{k-n}}{(k-n)! \prod_{i=1}^n \left(A_i^n + k - n\right)}$$

$$= (-1)^n t^{A_1^n} \sum_{k=0}^{\infty} \frac{(-t)^k}{k! \prod_{i=1}^n \left(A_i^n + k\right)}$$

$$= \frac{(-1)^n t^{A_1^n}}{\prod_{i=1}^n A_i^n} \sum_{k=0}^{\infty} \frac{(-t)^k}{k!} \cdot \prod_{i=1}^n \frac{(A_i^n)_k}{(A_i^n + 1)_k}$$

$$= \frac{(-1)^n t^{A_1^n}}{\prod_{i=1}^n A_i^n} \, {}_nF_n \left(\begin{matrix} A_1^n, & \ldots, & A_n^n \\ A_1^n + 1, & \ldots, & A_n^n + 1 \end{matrix} \middle| -t \right), \tag{4.19}$$

with $(a)_k = a \cdot (a+1) \cdots (a+k-1)$ and where ${}_nF_n$ denotes a generalized hypergeometric function, compare Appendix A.2. We apply Theorem A.5 which gives a Mellin-Barnes integral representation of generalized hypergeometric functions and obtain

$${}_nF_n \left(\begin{matrix} A_1^n, & \ldots, & A_n^n \\ A_1^n + 1, & \ldots, & A_n^n + 1 \end{matrix} \middle| -t \right)$$

$$= \prod_{i=1}^n A_i^n \cdot \frac{1}{2\pi i} \int_B t^z \frac{\Gamma(-z) \prod_{i=1}^n \Gamma\left(A_i^n + z\right)}{\prod_{i=1}^n \Gamma\left(A_i^n + 1 + z\right)} \, dz$$

$$= \prod_{i=1}^n A_i^n \cdot \frac{1}{2\pi i} \int_B t^z \frac{\Gamma(-z)}{\prod_{i=1}^n \left(A_i^n + z\right)} \, dz$$

$$= (-1)^n \prod_{i=1}^n A_i^n \cdot \frac{1}{2\pi i} \int_{B^*} t^{-z} \frac{\Gamma(z)}{\prod_{i=1}^n \left(z - A_i^n\right)} \, dz, \tag{4.20}$$

where B denotes Barnes' path of integration, that is, B is a line from $-i\infty$ to $+i\infty$ curving if necessary to put the poles of $\Gamma(-z)$ to the right of the path and those of $\Gamma(A_i^n + z)$, $i = 1, \ldots, n$, to the left of the path. Further B^* denotes a path that coincides with B when reflected at the imaginary axis. Additionally, let C_R for $R > 0$ be the closed path given in the figure below encircling $A_1^n, \ldots, A_n^n$ but not the poles of $\Gamma(z)$ in the positive direction:

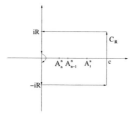

Residue calculus yields for $R > 0$

$$\frac{1}{2\pi i} \int_{C_R} t^{-z} \frac{\Gamma(z)}{\prod_{i=1}^n (z - A_i^n)} \, dz = \sum_{k=1}^n \frac{t^{-A_k^n} \Gamma(A_k^n)}{\prod_{i=1, i \neq k}^n (A_k^n - A_i^n)}$$

and hence

$$t^{A_1^n} \frac{1}{2\pi i} \int_{C_R} t^{-z} \frac{\Gamma(z)}{\prod_{i=1}^n (z - A_i^n)} \, dz = \sum_{k=1}^n \frac{t^{A_1^{k-1}} \Gamma(A_k^n)}{\prod_{i=1, i \neq k}^n (A_k^n - A_i^n)}$$

$$= \sum_{k=1}^n \frac{t^{A_1^{k-1}} \Gamma(A_k^n)}{\prod_{i=1}^{k-1} A_i^{k-1} \prod_{i=k+1}^n A_k^{i-1}} (-1)^{k-1}$$

$$= \sum_{k=1}^n \frac{t^{A_1^{k-1}} \Gamma(A_k^n)}{\prod_{i=1}^{k-1} A_i^{k-1} \prod_{i=k}^{n-1} A_k^i} (-1)^{k-1}$$

$$= \sum_{k=0}^{n-1} (-1)^k \frac{t^{A_1^k} \Gamma(A_{k+1}^n + 1)}{\prod_{i=1}^k A_i^k \prod_{i=k+1}^n A_{k+1}^i}. \qquad (4.21)$$

Denote by C_R^1 and C_R^2 the segments of C_R which run parallel to the real axis. We want to show that

$$\lim_{R \to \infty} \int_{C_R^i} t^{-z} \frac{\Gamma(z)}{\prod_{i=1}^n (z - A_i^n)} \, dz = 0$$

for $i = 1, 2$. For R sufficiently large we have

$$\left| t^{-z} \frac{\Gamma(z)}{\prod_{i=1}^n (z - A_i^n)} \right| \leq \left| t^{-z} \Gamma(z) \right|, \qquad n \in \mathbb{N}_0, z \in C_R^i, i = 1, 2,$$

wherefore we can restrict our arguments to the case $n = 0$. Due to Theorem A.2 in Section A.1 we obtain as $R \to \infty$

$$\ln \Gamma(z) = \left(z - \frac{1}{2} \right) \ln z - z + O(1), \qquad z \in C_R^i, i = 1, 2.$$

Therefore, as $R \to \infty$, the following holds for $z \in C_R^i$, $i = 1, 2$:

$$\ln\left(t^{-z}\Gamma(z)\right)$$

$$= -z\ln t + \left(z - \frac{1}{2}\right)\ln z - z + O(1)$$

$$= -(\Re(z) + i\Im(z))\ln t + \left(\Re(z) - \frac{1}{2} + i\Im(z)\right)(\ln|z| + i\arg(z))$$

$$\quad -(\Re(z) + i\Im(z)) + O(1)$$

$$= -\Re(z)(\ln t + 1) + \left(\Re(z) - \frac{1}{2}\right)\ln|z| - \Im(z)\arg(z) + PI + O(1),$$

where PI is a pure imaginary term. This implies for $z \in C_R^i$, $i = 1, 2$, and $R \to \infty$

$$t^{-z}\Gamma(z) = O\left(\exp\left\{-\Re(z)(\ln t + 1) + \left(\Re(z) - \frac{1}{2}\right)\ln|z| - \Im(z)\arg(z) + PI\right\}\right).$$

Now, for $z \in C_R^1$ (and for C_R^2 analogously) the following holds: $\Re(z) \in [0, c]$, $\arg(z) \in \left[\arg(c + iR), \frac{\pi}{2}\right]$, $\Im(z) = R$ and for $R \to \infty$ the term $-\Im(z)\arg(z)$ dominates $\left(\Re(z) - \frac{1}{2}\right)\ln|z|$.
Hence, we obtain

$$\lim_{R\to\infty}\left|\int_{C_R^i} t^{-z}\frac{\Gamma(z)}{\prod_{i=1}^n (z - A_i^n)}\,dz\right| \leq \lim_{R\to\infty} c\sup_{z\in C_R^i}\left|t^{-z}\frac{\Gamma(z)}{\prod_{i=1}^n (z - A_i^n)}\right|$$

$$\leq \lim_{R\to\infty} c\sup_{z\in C_R^i}\left|t^{-z}\Gamma(z)\right| = 0, \qquad i = 1, 2.$$

Altogether, we have for $R > 0$

$$\int_{C_R} t^{-z}\frac{\Gamma(z)}{\prod_{i=1}^n (z - A_i^n)}\,dz = \lim_{R\to\infty}\int_{C_R} t^{-z}\frac{\Gamma(z)}{\prod_{i=1}^n (z - A_i^n)}\,dz$$

$$= \int_{-B^*} t^{-z}\frac{\Gamma(z)}{\prod_{i=1}^n (z - A_i^n)}\,dz + \int_{c-i\infty}^{c+i\infty} t^{-z}\frac{\Gamma(z)}{\prod_{i=1}^n (z - A_i^n)}\,dz. \quad (4.22)$$

If we combine equations (4.18)-(4.22) we find

$$\sum_{k=0}^{\infty}(-1)^k \frac{t^{A_1^k}\,\Gamma\left(A_{k+1}^n + 1\right)}{\prod_{i=1}^k A_i^k \prod_{i=k+1}^n A_{k+1}^i}$$

$$\overset{(4.18),(4.19)}{=} \sum_{k=0}^{n-1}(-1)^k \frac{t^{A_1^k}\,\Gamma\left(A_{k+1}^n + 1\right)}{\prod_{i=1}^k A_i^k \prod_{i=k+1}^n A_{k+1}^i}$$

$$+ \frac{(-1)^n t^{A_1^n}}{\prod_{i=1}^n A_i^n} \, {}_nF_n\left(\begin{matrix} A_1^n, & \cdots, & A_n^n \\ A_1^n+1, & \ldots, & A_n^n+1 \end{matrix} \,\middle|\, -t \right)$$

$$\stackrel{(4.20)}{=} \sum_{k=0}^{n-1} (-1)^k \frac{t^{A_k^k} \, \Gamma\left(A_{k+1}^n+1\right)}{\prod_{i=1}^k A_i^k \prod_{i=k+1}^n A_{k+1}^i} + t^{A_1^n} \frac{1}{2\pi i} \int_{B^*} t^{-z} \frac{\Gamma(z)}{\prod_{i=1}^n (z-A_i^n)} \, dz$$

$$\stackrel{(4.21)}{=} t^{A_1^n} \frac{1}{2\pi i} \int_{C_R} t^{-z} \frac{\Gamma(z)}{\prod_{i=1}^n (z-A_i^n)} \, dz + t^{A_1^n} \frac{1}{2\pi i} \int_{B^*} t^{-z} \frac{\Gamma(z)}{\prod_{i=1}^n (z-A_i^n)} \, dz$$

$$\stackrel{(4.22)}{=} t^{A_1^n} \frac{1}{2\pi i} \int_{c-i\infty}^{c+i\infty} t^{-z} \frac{\Gamma(z)}{\prod_{i=1}^n (z-A_i^n)} \, dz$$

for $R > 0$ which proves the proposition. ∎

Note that the assumptions of Proposition 54 hold especially for sequences $\{a_i\}_{i\in\mathbb{N}} \subset (-1,\infty)$ eventually constant exceeding -1.

Moreover, it is probably also possible to achieve a series representation similar to (4.17) if we admit the values $A_1^n,\ldots,A_n^n$ to be equal. However, this representation might be fairly complicated.

To summarize, given $n \in \mathbb{N}_0$, $a > -1$ and a sequence $\{a_i\}_{i\in\mathbb{N}}$ such that $a_i = a$ for $i > n$, we have found three different representations of Φ_λ^n for $\lambda > 0$, that are

$$\Phi_\lambda^n(t \,|\, a_1,\ldots,a_n; a) = \int_t^\infty \int_{s_1}^\infty \cdots \int_{s_{n-1}}^\infty \prod_{i=1}^n s_i^{a_i} \cdot e^{-\lambda s_n^{a+1}} \, ds_n \cdots ds_1$$

$$= \left(t^{a+1}\right)^{A_1^n} (a+1)^{-n} \cdot \frac{1}{2\pi i} \int_{c-i\infty}^{c+i\infty} \left(\lambda t^{a+1}\right)^{-z} \frac{\Gamma(z)}{\prod_{i=1}^n (z-A_i^n)} \, dz$$

$$= \lambda^{-A_1^n}(a+1)^{-n} \sum_{k=0}^\infty (-1)^k \frac{\left(\lambda t^{a+1}\right)^{A_1^k} \Gamma\left(A_{k+1}^n+1\right)}{\prod_{i=1}^k A_i^k \prod_{i=k+1}^n A_{k+1}^i}, \qquad t > 0, \qquad (4.23)$$

where the last equation holds for $A_1^n,\ldots,A_n^n$ mutually distinct, none of it being 0 or a negative integer, only. Let us return to the generalized order statistic process N considered at the beginning with respect to a sequence $\{\alpha_i\}_{i\in\mathbb{N}}$ such that $\alpha_i = \alpha$ for $i > n^*$, some $n^* \in \mathbb{N}$ and $\alpha > -1$, and with the densities

$$f_{T_1,\ldots,T_n}(t_1,\ldots,t_n)$$

$$= \frac{\lambda^{\frac{\gamma_{n^*}}{\alpha+1}} \prod_{i=1}^{n^*} \gamma_i}{\Gamma\left(\frac{\gamma_{n^*}}{\alpha+1}+1\right)} \prod_{i=1}^n t_i^{\alpha_i} \int_{t_n}^\infty \int_{s_{n+1}}^\infty \cdots \int_{s_{n^*-1}}^\infty \prod_{i=n+1}^{n^*} s_i^{\alpha_i} \cdot e^{-\lambda s_{n^*}^{\alpha+1}} \, ds_{n^*} \cdots ds_{n+1}$$

for $n < n^*$ and $0 < t_1 \leq \cdots \leq t_n$. In terms of $A_i^k = \sum_{j=i}^k \frac{\alpha_i+1}{\alpha+1}$, $i, k \in \mathbb{N}$, Propositions 53 and 54 yield the following representations for the densities of $n \leq n^*$ successive occurrence times:

$$f_{T_1,\dots,T_n}(t_1,\dots,t_n)$$

$$= \frac{\lambda^{\frac{\gamma_{n^*}}{\alpha+1}} \prod_{i=1}^{n^*} \gamma_i}{\Gamma\left(\frac{\gamma_{n^*}}{\alpha+1}+1\right)} \prod_{i=1}^{n} t_i^{\alpha_i} \cdot \Phi_\lambda^{n^*-n}(t_n \mid \alpha_{n+1},\dots,\alpha_{n^*};\alpha) \tag{4.24}$$

$$= \frac{\lambda^{A_1^{n^*}} \prod_{i=1}^{n^*} A_1^i}{\Gamma\left(A_1^{n^*}+1\right)}(\alpha+1)^n \prod_{i=1}^{n} t_i^{\alpha_i} \frac{\left(t_n^{\alpha+1}\right)^{A_{n+1}^{n^*}}}{2\pi\mathrm{i}} \int_{c-i\infty}^{c+i\infty} \left(\lambda t_n^{\alpha+1}\right)^{-z} \frac{\Gamma(z)}{\prod_{i=n+1}^{n^*} \left(z-A_i^{n^*}\right)} dz$$

$$= \frac{\lambda^{A_1^n} \prod_{i=1}^{n^*} A_1^i}{\Gamma\left(A_1^{n^*}+1\right)}(\alpha+1)^n \prod_{i=1}^{n} t_i^{\alpha_i} \sum_{k=0}^{\infty}(-1)^k \frac{\left(\lambda t_n^{\alpha+1}\right)^{A_{n+k}^{n^*}} \Gamma\left(A_{n+k+1}^{n^*}+1\right)}{\prod_{i=n+1}^{n+k} A_i^{n+k} \prod_{i=n+k+1}^{n^*} A_{n+k+1}^i}, \tag{4.25}$$

where $0 < t_1 \le \cdots \le t_n$ and where representation (4.25) again holds if $A_{n+1}^{n^*},\dots,A_{n^*}^{n^*}$ are mutually distinct and $\notin -\mathbb{N}_0$, only.

Note that with formula (4.25) we finally achieved a representation departing from which it seems to be manageable to consider the limit $n^* \to \infty$. Corresponding studies can be found in Section 5.3.

Example 55: For a parametrizing sequence such that $\alpha_i = 0$ for $i > n^*$ and $\alpha_i = -\frac{1}{2}$ else for some $n^* \in \mathbb{N}$ we find

$$A_{i+1}^j = \;=\; \begin{cases} \frac{j-i}{2} & \text{if } 0 \le i < j \le n^*, \\ j-n^*+\frac{n^*-i}{2} & \text{if } 0 \le i < n^* < j, \\ j-i & \text{if } n^* \le i < j, \\ 0 & \text{else.} \end{cases}$$

Since e.g. for $n \le n^* \le n+k$ we obtain

$$\prod_{i=n+1}^{n+k} A_i^{n+k} = \prod_{i=n+1}^{n^*} A_i^{n+k} \prod_{i=n^*+1}^{n+k} A_i^{n+k}$$

$$= \prod_{i=n+1}^{n^*} \left(n+k-n^*+\frac{n^*-(i-1)}{2}\right) \prod_{i=n^*+1}^{n+k} \left(n+k-(i-1)\right)$$

$$= \frac{1}{2^{n^*-n}} \prod_{i=n+1}^{n^*} \left(2(n+k)-n^*-i+1\right) \cdot (n+k-n^*)!$$

$$= \frac{1}{2^{n^*-n}} \left(2(n+k-n^*)+1\right)_{n^*-n} \cdot (n+k-n^*)!,$$

where $(x)_n = x \cdot (x+1) \cdots (x+n-1)$ for $n \in \mathbb{N}$ and $(x)_0 = 1$, in our example, (4.25) coincides with

$$f_{T_1,\ldots,T_n}(t_1,\ldots,t_n)$$

$$= \frac{\frac{n^*!}{2^{n^*}}\sqrt{\lambda}^n}{\Gamma\left(\frac{n^*}{2}+1\right)} \prod_{i=1}^{n} t_i^{-\frac{1}{2}} \left(\sum_{k=0}^{n^*-n-1} (-1)^k \frac{\sqrt{\lambda t_n}^{-k}\Gamma\left(\frac{n^*-n-k}{2}+1\right)}{\frac{k!}{2^k}\cdot\frac{(n^*-n-k)!}{2^{n^*-n-k}}} \right.$$

$$\left. + \sum_{k=n^*-n}^{\infty} (-1)^k \frac{(\lambda t_n)^{n+k-n^*+\frac{n^*-n}{2}}}{\frac{(2(n+k-n^*)+1)_{n^*-n}}{2^{n^*-n}}(n+k-n^*)!} \right)$$

$$= \frac{n^*!\sqrt{\lambda}^n}{2^n\Gamma\left(\frac{n^*}{2}+1\right)} \prod_{i=1}^{n} t_i^{-\frac{1}{2}} \left(\sum_{k=0}^{n^*-n-1} (-1)^k \frac{\sqrt{\lambda t_n}^{-k}\Gamma\left(\frac{n^*-n-k}{2}+1\right)}{k!(n^*-n-k)!} \right.$$

$$\left. + \sum_{k=0}^{\infty} (-1)^{n^*+k-n} \frac{(\lambda t_n)^{k+\frac{n^*-n}{2}}}{(2k+1)_{n^*-n}k!} \right), \qquad n \le n^*, 0 < t_1 \le \cdots \le t_n.$$

Looking closely, parts of the last formula resemble the following densities corresponding to a constant parametrizing sequence such that $\alpha_i = -\frac{1}{2}$, $i \in \mathbb{N}$:

$$f_{T_1,\ldots,T_n}(t_1,\ldots,t_n) = \frac{\sqrt{\lambda}^n}{2^n\sqrt{t_1\cdots t_n}}e^{-\sqrt{\lambda t_n}}, \qquad n \in \mathbb{N}, 0 < t_1 \le \cdots \le t_n,$$

see Theorem 40. This encourages our attempt to obtain generators considering the limit $n^* \to \infty$.   □

Chapter 5

Generators of generalized order statistic processes

Until now we have well studied GOS-processes w. r. t. eventually constant sequences. This section is finally dedicated to present some processes satisfying the generalized order statistic property w. r. t. more general parametrizing sequences. Hence, our aim is to specify a family $\{\varphi_n\}_{n\in\mathbb{N}_0}$ and thus the distribution of corresponding GOS-processes. We will concentrate on nonnegative, increasing convergent and periodic sequences. In order to do so, we introduce the formal concept of a generator:

To begin with, recall Theorem 32 where we stated that $n \in \mathbb{N}$ successive occurrence times $T_1, \ldots, T_n$ of a GOS-process w. r. t. a sequence $\{\alpha_i\}_{i\in\mathbb{N}}$ follow the density

$$f_{T_1,\ldots,T_n}(t_1,\ldots,t_n) = \prod_{i=1}^{n} t_i^{\alpha_i} \varphi_n(t_n), \qquad 0 < t_1 \leq \cdots \leq t_n, \qquad (5.1)$$

for some family of functions $\{\varphi_n\}_{n\in\mathbb{N}}$, provided the process jumps at least once. The functions $\{\varphi_n\}_{n\in\mathbb{N}_0}$, where $\varphi_0(t) = \int_t^\infty s^{\alpha_1}\varphi_1(s)ds$, $t > 0$, are interrelated by the differential equation (3.11):

$$-\varphi_{n+1}(t)\, t^{\alpha_{n+1}} = \dot{\varphi}_n(t), \qquad n \in \mathbb{N}_0, t > 0.$$

Thus, the family $\{\varphi_n\}_{n\in\mathbb{N}_0}$ and consequently the distribution of the GOS-process N w. r. t. $\{\alpha_i\}_{i\in\mathbb{N}}$ (up to the probability $P(T_1 < \infty)$) are completely determined by a single smooth function φ_0 and the above recursive system of differential equations. This fact motivates the following definition:

Definition 56: *Let $\{\alpha_i\}_{i\in\mathbb{N}}$ be a real sequence and $\varphi_0 : (0,\infty) \to [0,\infty)$ an infinitely often differentiable function such that $\lim_{t\to 0}\varphi_0(t) = 1$. We call φ_0 a* **generator** *with respect to $\{\alpha_i\}_{i\in\mathbb{N}}$ if the family $\{\varphi_n\}_{n\in\mathbb{N}}$ which is recursively defined by*

$$-\varphi_{n+1}(t)\, t^{\alpha_{n+1}} = \dot{\varphi}_n(t), \qquad n \in \mathbb{N}_0, t > 0, \qquad (5.2)$$

verifies

$$\varphi_n(t) \geq 0, \qquad n \in \mathbb{N}_0, t > 0, \qquad (5.3)$$

$$\lim_{t\to\infty} \varphi_n(t) = 0, \qquad n \in \mathbb{N}_0. \qquad (5.4)$$

For technical reasons put $\varphi_0(0) = 1$.

A function φ_0 which is a generator naturally corresponds to a GOS-process (jumping almost surely) since the functions $\{f_n\}_{n \in \mathbb{N}}$ given by

$$f_n(t_1, \ldots, t_n) = \prod_{i=1}^{n} t_i^{\alpha_i} \, \varphi_n(t) \cdot \mathbb{1}_{K_n}(t_1, \ldots, t_n), \qquad n \in \mathbb{N}, \, t_1, \ldots, t_n \in \mathbb{R},$$

form a projective family of probability densities and since then Kolmogorov's existence theorem assures the existence of a process whose successive occurrence times T_i, $i \in \mathbb{N}$, are such that $T_1, \ldots, T_n$ follow the joint density given by f_n, $n \in \mathbb{N}$. Thereby, condition (5.2) assures the projectivity of the densities, (5.3) their nonnegativity, (5.4) and the fact that $\lim_{t \to 0} \varphi_0(t) = 1$ assure the integrability and normalization of the densities.

Further GOS-processes can be constructed such that the distribution $P_{T_1, \ldots, T_n}$ of successive occurrence times of such a process satisfies

$$P_{T_1, \ldots, T_n} = p_{<\infty} f_n \odot \bar{\ell}^n + (1 - p_{<\infty}) \, \delta_{(\infty, \ldots, \infty)}, \qquad n \in \mathbb{N},$$

for some $p_{<\infty}$ such that $p_{<\infty} = P(T_1 < \infty)$, compare Lemma 29. Then, the generator has the following probabilistic interpretation:

$$\varphi_0(t) = P(T_1 > t | T_1 < \infty) = P(N_t = 0 | T_1 < \infty), \qquad t \geq 0,$$

hence φ_0 corresponds to the survival function of the first occurrence time T_1, provided that a jump occurs.

Up to now, we cannot exclude the possibility that a thus constructed GOS-process explodes with positive probability. However, in terms of the generator, corresponding GOS-processes do not explode almost surely if and only if

$$\sum_{k=0}^{\infty} \frac{t^{\gamma k}}{\prod_{i=1}^{k} \gamma_i} \varphi_k(t) = 1, \qquad \text{for all } t > 0. \tag{5.5}$$

That is the case if and only if $\sum_{i=0}^{\infty} \frac{1}{\gamma_i}$ diverges, compare Proposition 36 and (3.25).

Altogether, a generator determines the distribution of a GOS-process up to the probability $P(T_1 < \infty)$. The question whether GOS-processes w.r.t. a certain parametrizing sequence exist or not is therefore equivalent to whether or not we can find a generator w.r.t. the sequence.

In the case of mixed Poisson processes, a similar construction of a subclass of mixed Poisson processes based on a generating function has been proposed by Hofmann (1955) in the context of insurance mathematics. Aiming at distributions which can be calculated relatively simply, Hofmann in principal chooses generators of the form

$$\varphi_0(t) = e^{-q\tau(t)},$$

where $q > 0$ and where $\tau : [0, \infty) \to [0, \infty)$ is a differentiable function with $\tau(0) = 0$ whose first derivative $\dot{\tau}$ satisfies $\dot{\tau}(t) = \left(\frac{1}{1+ct}\right)^a$ for some $c > 0$ and $a > 0$. Corresponding mixing distributions include Gamma- or inverse Gaussian distributions, compare e. g. Walhin and Paris (1999).

The outline for the rest of the chapter is as follows: In Section 5.1 we deduce generators w. r. t. nonnegative parametrizing sequences. Its results will be exploited by an excursus where we simulate paths of some GOS-processes. Section 5.3 presents generators with respect to in principle increasing, convergent sequences, Section 5.4 shows a method how to deduce generators w. r. t. periodic sequences. One remarkable issue concerning all these results is, that generators w. r. t. a given sequence turn out to be nonunique. Therefore, in Section 5.5 we study the relation of different generators w. r. t. one and the same parametrizing sequence.

5.1 Generators with respect to nonnegative sequences

Let us begin with the case of nonnegative parametrizing sequences. The following proposition shows that the Laplace transform of an arbitrary probability measure on $(0, \infty)$ is a generator with respect to such a sequence:

Theorem 57: Let $\{\alpha_i\}_{i \in \mathbb{N}}$ be a nonnegative sequence and W a probability distribution with support $(0, \infty)$. Then

$$\varphi_0(t) = \int_{(0,\infty)} e^{-\lambda t} \, dW(\lambda), \qquad t \geq 0, \tag{5.6}$$

is a generator with respect to $\{\alpha_i\}_{i \in \mathbb{N}}$.

Proof: Consider a nonnegative sequence $\{\alpha_i\}_{i \in \mathbb{N}}$ and a probability distribution W with support $(0, \infty)$. Notice that $\varphi_0(t) = \int_{(0,\infty)} e^{-\lambda t} \, dW(\lambda)$, $t > 0$, is positive and such that $\lim_{t \to 0} \varphi_0(t) = 1$ and $\lim_{t \to \infty} \varphi_0(t) = 0$.

We will prove by induction that for $n \in \mathbb{N}$ there exists a polynomial $g_n(x) = \sum_{k=1}^{n} c_k^n x^k$ with nonnegative coefficients $c_1^n, \ldots, c_n^n$ such that

$$\varphi_n(t) = t^{-\gamma_n} \int_{(0,\infty)} g_n(\lambda t) \, e^{-\lambda t} \, dW(\lambda), \qquad t > 0. \tag{5.7}$$

This in consequence implies condition (5.3) and further (5.4) by dominated convergence.

For $n = 1$ we have

$$\varphi_1(t) = -\frac{\dot{\varphi}_0(t)}{t^{\alpha_1}} = \frac{\int_{(0,\infty)} \lambda e^{-\lambda t} \, dW(\lambda)}{t^{\alpha_1}} = \frac{\int_{(0,\infty)} g_1(\lambda t) e^{-\lambda t} \, dW(\lambda)}{t^{\gamma_1}}, \qquad t > 0,$$

where $g_1(x) = x$ for $x > 0$. Thus, φ_1 is of the type specified in equation (5.7). Now, for some $n \in \mathbb{N}$, let

$$\varphi_n(t) = t^{-\gamma_n} \int_{(0,\infty)} g_n(\lambda t)\, e^{-\lambda t}\, dW(\lambda), \qquad t > 0,$$

be verified for some $g_n(x) = \sum_{k=1}^{n} c_k^n x^k$, $x > 0$, with nonnegative coefficients c_k^n, $k = 1,\ldots,n$. Then

$$\varphi_{n+1}(t) = -\frac{\dot\varphi_n(t)}{t^{\alpha_{n+1}}} = -\frac{\frac{d}{dt}\, t^{-\gamma_n} \int_{(0,\infty)} g_n(\lambda t)\, e^{-\lambda t}\, dW(\lambda)}{t^{\alpha_{n+1}}}$$

$$\overset{\text{Prop.C.6}}{=} -\frac{\int_{(0,\infty)} \frac{d}{dt}\left(t^{-\gamma_n} \sum_{k=1}^{n} c_k^n (\lambda t)^k\, e^{-\lambda t}\right) dW(\lambda)}{t^{\alpha_{n+1}}}$$

$$= -\frac{\int_{(0,\infty)} \frac{d}{dt}\left(e^{-\lambda t} \sum_{k=1}^{n} c_k^n \lambda^k t^{k-\gamma_n}\right) dW(\lambda)}{t^{\alpha_{n+1}}}$$

$$= -\frac{\int_{(0,\infty)} e^{-\lambda t} \sum_{k=1}^{n} c_k^n \lambda^k (k-\gamma_n) t^{k-\gamma_n-1} - \lambda e^{-\lambda t} \sum_{k=1}^{n} c_k^n \lambda^k t^{k-\gamma_n} dW(\lambda)}{t^{\alpha_{n+1}}}$$

$$= \frac{\int_{(0,\infty)} \left(\sum_{k=1}^{n} c_k^n (\gamma_n - k)(\lambda t)^k + \sum_{k=1}^{n} c_k^n (\lambda t)^{k+1}\right) e^{-\lambda t} dW(\lambda)}{t^{\gamma_{n+1}}}$$

$$= t^{-\gamma_{n+1}} \int_{(0,\infty)} g_{n+1}(\lambda t)\, e^{-\lambda t}\, dW(\lambda), \qquad t > 0,$$

where

$$g_{n+1}(x) = c_1^n(\gamma_n - 1)x + \sum_{k=2}^{n} \left(c_k^n(\gamma_n - k) + c_{k-1}^n\right) x^k + c_n^n x^{n+1}, \qquad x > 0,$$

has the required form. Since $\alpha_i \geq 0$ for $i \in \mathbb{N}$ implies $\gamma_i \geq i$ for $i \in \mathbb{N}$ the polynomial g_{n+1} has nonnegative coefficients $c_0^{n+1},\ldots,c_n^{n+1}$. ∎

For the generator $\varphi_0(t) = \int_{(0,\infty)} e^{-\lambda t}\, dW(\lambda)$, $t \geq 0$, with respect to a nonnegative sequence $\{\alpha_i\}_{i\in\mathbb{N}}$, we specify the first three functions of the family $\{\varphi_n\}_{n\in\mathbb{N}}$:

$$\varphi_1(t) = t^{-\gamma_1} \int_{(0,\infty)} \lambda t e^{-\lambda t}\, dW(\lambda),$$

$$\varphi_2(t) = t^{-\gamma_2} \int_{(0,\infty)} \left((\lambda t)^2 + \alpha_1 \lambda t\right) e^{-\lambda t}\, dW(\lambda),$$

$$\varphi_3(t) = t^{-\gamma_3} \int_{(0,\infty)} \left((\lambda t)^3 + (2\alpha_1 + \alpha_2)(\lambda t)^2 + \alpha_1(\alpha_1 + \alpha_2 + 1)\lambda t\right) e^{-\lambda t}\, dW(\lambda),$$

where $t > 0$.

Note that the generators given by formula (5.6) do not include all possible generators w. r. t. nonnegative sequences:

Look for instance at $\alpha_1 = 1$ and $\alpha_i = 0$ for $i > 1$. Due to Theorem 40 we can specify φ_1 up to a mixture and hence in this case all generators can be represented by

$$\varphi_0(t) \;=\; \int_t^\infty \int_{(0,\infty)} s\lambda^2 e^{-\lambda s}\, dW(\lambda)ds \;=\; \int_{(0,\infty)} (\lambda t + 1)e^{-\lambda t}\, dW(\lambda), \qquad t \geq 0,$$

for some probability distribution W on $(0,\infty)$. Choose especially $\varphi_0(t) = (\tilde\lambda t + 1)e^{-\tilde\lambda t}$ for some $\tilde\lambda > 0$ as a generator. Since $\varphi_0(t) = (\tilde\lambda t + 1)e^{-\tilde\lambda t}$ is not completely monotone, because its second derivative changes its sign at $t = \tilde\lambda^{-1}$,

$$\dot\varphi_0(t) \;=\; \tilde\lambda e^{-\tilde\lambda t} - (\tilde\lambda t + 1)\tilde\lambda e^{-\tilde\lambda t} \;=\; -\tilde\lambda^2 t e^{-\tilde\lambda t},$$

$$\ddot\varphi_0(t) \;=\; \tilde\lambda^3 t e^{-\tilde\lambda t} - \tilde\lambda^2 e^{-\tilde\lambda t},$$

the chosen generator is no Laplace transform, that is, there cannot exist a distribution W such that $(\tilde\lambda t + 1)e^{-\tilde\lambda t} = \int_{(0,\infty)} e^{-\lambda t}\, dW(\lambda)$.

Lastly, note that Theorem 57 can be extended to parametrizing sequences admitting a lower bound:

Corollary 58: Let $\alpha > -1$, $\alpha_i \geq \alpha$, $i \in \mathbb{N}$, and W a probability distribution on $(0,\infty)$. Then

$$\varphi_0(t) \;=\; \int_{(0,\infty)} e^{-\lambda t^{\alpha+1}}\, dW(\lambda), \qquad t \geq 0, \tag{5.9}$$

is a generator with respect to $\{\alpha_i\}_{i\in\mathbb{N}}$.

Proof: Consider $\tilde\alpha_i = \frac{\alpha_i+1}{\alpha+1} - 1$, $i \in \mathbb{N}$. Then $\tilde\alpha_i \geq 0$ for $i \in \mathbb{N}$ and due to Thm. 57

$$\tilde\varphi_0(t) \;=\; \int_{(0,\infty)} e^{-\lambda t}dW(\lambda), \qquad t > 0,$$

is a generator w. r. t. $\{\tilde\alpha_i\}_{i\in\mathbb{N}}$ for arbitrary probability distributions W on $(0,\infty)$. The probabilistic meaning of $\tilde\varphi_0$ is

$$\tilde\varphi_0(t) \;=\; P\left(\tilde T_1 > t\right), \qquad t > 0,$$

where $\tilde T_1$ is the first occurrence time of a GOS-process $\tilde N$ w. r. t. $\{\tilde\alpha_i\}_{i\in\mathbb{N}}$ jumping almost surely.

According to Proposition 34, a time transformation $t \mapsto t^{\alpha+1}$ of $\tilde N$ yields a GOS-process N w. r. t. $\{\alpha_i\}_{i\in\mathbb{N}}$ whose first occurrence time T_1 satisfies $T_1^{\alpha+1} = \tilde T_1$. Therefore,

$$P(T_1 > t) = P\left(\tilde T_1 > t^{\alpha+1}\right) = \int_{(0,\infty)} e^{-\lambda t^{\alpha+1}}dW(\lambda), \qquad t > 0,$$

and $\int_{(0,\infty)} e^{-\lambda t^{\alpha+1}}dW(\lambda)$ is the generator corresponding to N. $\blacksquare$

5.2   Excursus – Simulation of generalized order statistic processes

From the previous section we can conclude that w. r. t. a nonnegative sequence $\{\alpha_i\}_{i \in \mathbb{N}}$ and $n^* \in \mathbb{N}$ there exists a GOS-process such that the corresponding function φ_{n^*} equals $\varphi_{n^*}(t) = \frac{\prod_{i=1}^{n^*} \gamma_i}{\Gamma(\gamma_{n^*}+1)} e^{-t}$, $t > 0$:

Firstly, note that $\prod_{i=1}^{n^*} t_i^{\alpha_i} \varphi_{n^*}(t_{n^*}) = \frac{\prod_{i=1}^{n^*} \gamma_i t_i^{\alpha_i}}{\Gamma(\gamma_{n^*}+1)} e^{-t_{n^*}}$ is a density on K_{n^*}. Further, arguments similar to those in the proof of Theorem 57 show that the functions φ_n for $n > n^*$, recursively obtained by (5.2), verify conditions (5.3) and (5.4). Hence, φ_{n^*} corresponds to a well defined GOS-process with densities $f_{T_1,\dots,T_n}(t_1,\dots,t_n) = \prod_{i=1}^{n} t_i^{\alpha_i} \varphi_n(t_n)$ on K_n for $n \geq n^*$.

In the sequel we want to simulate such GOS-processes w. r. t. three different nonnegative sequences until the n^*-th occurrence time:

The algorithm to apply is based on the results of Section 3.5.3, saying that for a GOS-process the ratios of occurrence times $\frac{T_i}{T_{i+1}}$ for $i = 1,\dots,n^*-1$ and T_{n^*} are independent with $\frac{T_i}{T_{i+1}} \sim B(\gamma_i, 1)$. Moreover, T_{n^*} is $\Gamma(1, \gamma_{n^*})$-distributed due to (3.13) and the chosen φ_{n^*}. Thus, the MATLAB procedure which provides the subsequent examples and which can be found in Appendix E starts with the simulation of a $\Gamma(1, \gamma_{n^*})$-distributed random variable as realization of T_{n^*} using the algorithm described in Appendix D. It requires $\gamma_{n^*} > 1$ why we choose $n^* > 1$. Then, for $i = 1,\dots,n^*-1$, a random variable B_i which is Beta distributed with parameters γ_i and 1 and which is independent of T_{n^*} and B_j for $j \neq i$ is simulated using the inverse distribution function method. Finally, for $i = 1,\dots,n^*-1$, T_i is recursively calculated via $T_i = B_i \cdot T_{i+1}$.

Example 59: The following plots show the path of a GOS-process w. r. t. the sequence such that $\alpha_{2i-1} = 0$, $\alpha_{2i} = 1$ for $i \in \mathbb{N}$ and corresponding to $\varphi_{4000}(t) = \frac{\prod_{i=1}^{4000} \gamma_i}{\Gamma(\gamma_{4000}+1)} e^{-t}$ for $t > 0$, i. e. we choose $n^* = 4000$. The left figure displays the path until the 40th occurrence time, the right figure until the 4000th:

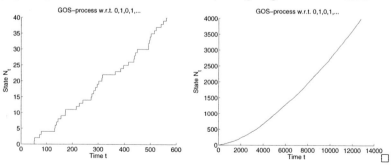

Example 60: For a GOS-process w. r. t. the sequence $\alpha_i = i - 1$ for $i \in \mathbb{N}$ and such that $\varphi_{4000}(t) = \frac{\prod_{i=1}^{4000} \gamma_i}{\Gamma(\gamma_{4000}+1)} e^{-t}$ for $t > 0$, we find the following realization shown on the left side until the 40th and on the right side until the 4000th occurrence time:

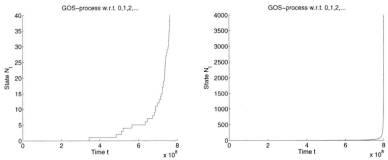

Note that the simulated path apparently explodes, i. e. infinitely many jumps occur in finite time. Therefore, this example reflects Proposition 36 stating that explosion takes place if $\sum_{i=1}^{\infty} \frac{1}{\gamma_i}$ converges, which in our case corresponds to the convergence of $\sum_{i=1}^{\infty} \frac{2}{i(i+1)}$ as $\gamma_i = \frac{i(i+1)}{2}$ for $i \in \mathbb{N}$. $\qquad\square$

Example 61: Our last example shows several paths of a GOS-process w. r. t. the sequence $\alpha_i = 1 - 0.5^{i-1}$ for $i \in \mathbb{N}$ and such that $\varphi_{4000}(t) = \frac{\prod_{i=1}^{4000} \gamma_i}{\Gamma(\gamma_{4000}+1)} e^{-t}$ for $t > 0$. We find the following 100 realizations plotted until the 40th occurrence time on the left side and until the 4000th on the right side:

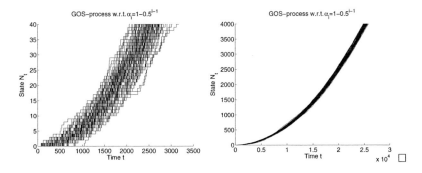

All the above realizations evidently exhibit some kind of regular asymptotic behavior, reflecting also the result of Corollary 52, which, in the light of this section's examples, probably can be extended to a more general setting.

5.3   Generators with respect to increasing convergent sequences

We construct generators with respect to mainly increasing convergent sequences. Essentially, the results of this section are based on those of Section 4.5 where we deduced different representations in the case of GOS-processes whose parametrizing sequences are constant from some index n^* on. The principal idea now is to consider more and more fluctuating sequences. We let n^* tend to infinity and draw conclusions by means of this limit considerations about generators in a more general setting.

Let $\{\alpha_i\}_{i\in\mathbb{N}} \subset (-1,\infty)$ be a sequence such that $\alpha_i = \alpha$, $i > n^*$, for some $n^* \in \mathbb{N}$ and $\alpha > -1$. Recall the representation given in Section 4.5 of the first occurrence time's density of a generalized order statistic process N w.r.t. $\{\alpha_i\}_{i\in\mathbb{N}}$ provided that at least one jump occurs. Neglecting again the possible presence of a mixing distribution W we may thus write

$$f_{T_1}(t) \;=\; \frac{\lambda^{\frac{\gamma_{n^*}}{\alpha+1}}\prod_{i=1}^{n^*}\gamma_i}{\Gamma\left(\frac{\gamma_{n^*}}{\alpha+1}+1\right)}t^{\alpha_1} \cdot \Phi_\lambda^{n^*-1}(t\,|\,\alpha_2,\ldots,\alpha_{n^*};\alpha), \qquad t>0,$$

for some $\lambda > 0$ (fixed for the rest of this section). Thereby $\Phi_\lambda^{n^*-1}$ is one of the functions defined in (4.14):

$$\Phi_\lambda^n(t\,|\,\alpha_1,\ldots,\alpha_n;\alpha) \;=\; \int_t^\infty\int_{s_1}^\infty\cdots\int_{s_{n-1}}^\infty \prod_{i=1}^n s_i^{\alpha_i}\cdot e^{-\lambda s_n^{\alpha+1}}\,ds_n\cdots ds_1 \qquad (5.10)$$

for $n \in \mathbb{N}$ resp. $\Phi_\lambda^0(t|\alpha) = e^{-\lambda t^{\alpha+1}}$, $t > 0$. The generator φ_0 corresponding to N verifies

$$\varphi_0(t) \;=\; P(T_1 > t|T_1 < \infty) \;=\; \int_t^\infty f_{T_1}(s)ds$$

$$= \;\frac{\lambda^{\frac{\gamma_{n^*}}{\alpha+1}}\prod_{i=1}^{n^*}\gamma_i}{\Gamma\left(\frac{\gamma_{n^*}}{\alpha+1}+1\right)}\int_t^\infty s^{\alpha_1}\Phi_\lambda^{n^*-1}(s\,|\,\alpha_2,\ldots,\alpha_{n^*};\alpha)\,ds$$

$$\stackrel{(4.15)}{=} \;\frac{\lambda^{\frac{\gamma_{n^*}}{\alpha+1}}\prod_{i=1}^{n^*}\gamma_i}{\Gamma\left(\frac{\gamma_{n^*}}{\alpha+1}+1\right)}\Phi_\lambda^{n^*}(t\,|\,\alpha_1,\alpha_2,\ldots,\alpha_{n^*};\alpha), \qquad t\geq 0. \qquad (5.11)$$

In the sequel we write $\varphi_0^{n^*}(t|\alpha_1,\ldots,\alpha_{n^*};\alpha)$ instead of $\varphi_0(t)$ to emphasize its dependence on n^*, $\alpha_1,\ldots,\alpha_{n^*}$ and α.

To find a candidate for a generator with respect to a more general sequence $\{\alpha_i\}_{i\in\mathbb{N}}$ we let n^* tend to infinity in equation (5.11). Thereby we exploit the series representation of $\Phi_\lambda^{n^*}$ given in Section 4.5

$$\Phi_\lambda^{n^*}(t\,|\,\alpha_1,\alpha_2,\ldots,\alpha_{n^*};\alpha) \;\stackrel{(4.23)}{=}\; \lambda^{-A_1^{n^*}}(\alpha+1)^{-n^*}\sum_{k=0}^\infty(-1)^k\frac{(\lambda t^{\alpha+1})^{A_1^k}\,\Gamma\left(A_{k+1}^{n^*}+1\right)}{\prod_{i=1}^k A_i^k\,\prod_{i=k+1}^{n^*}A_{k+1}^i},$$

where $A_i^k = \sum_{j=i}^k \frac{\alpha_j+1}{\alpha+1}$ for $i, k \in \mathbb{N}$. Then the formal limit of $\varphi_0^{n^*}$ can be written as

$$\varphi_0^\infty(t|\alpha_1, \alpha_2, \ldots; \alpha) \;=\; \lim_{n^* \to \infty} \varphi_0^{n^*}(t|\alpha_1, \ldots, \alpha_{n^*}; \alpha)$$

$$= \; \lim_{n^* \to \infty} \frac{\lambda^{\frac{\gamma_{n^*}}{\alpha+1}} \prod_{i=1}^{n^*} \gamma_i}{\Gamma\left(\frac{\gamma_{n^*}}{\alpha+1} + 1\right)} \Phi_\lambda^{n^*}(t \,|\, \alpha_1, \alpha_2, \ldots, \alpha_{n^*}; \alpha)$$

$$\overset{(4.23)}{=} \; \lim_{n^* \to \infty} \frac{\lambda^{\frac{\gamma_{n^*}}{\alpha+1}} \prod_{i=1}^{n^*} \gamma_i}{\Gamma\left(\frac{\gamma_{n^*}}{\alpha+1} + 1\right)} \lambda^{-A_1^{n^*}} (\alpha+1)^{-n^*} \sum_{k=0}^\infty (-1)^k \frac{(\lambda t^{\alpha+1})^{A_1^k} \, \Gamma\left(A_{k+1}^{n^*} + 1\right)}{\prod_{i=1}^k A_i^k \prod_{i=k+1}^{n^*} A_{k+1}^i}$$

$$= \; \lim_{n^* \to \infty} \frac{\prod_{i=1}^{n^*} A_1^i}{\Gamma\left(A_1^{n^*} + 1\right)} \sum_{k=0}^\infty (-1)^k \frac{(\lambda t^{\alpha+1})^{A_1^k} \, \Gamma\left(A_{k+1}^{n^*} + 1\right)}{\prod_{i=1}^k A_i^k \prod_{i=k+1}^{n^*} A_{k+1}^i} \tag{5.12}$$

$$= \; \lim_{n^* \to \infty} \frac{\prod_{i=1}^{n^*} A_1^i}{\Gamma\left(A_1^{n^*} + 1\right)} \sum_{k=0}^\infty (-1)^k \frac{(\lambda t^{\alpha+1})^{A_1^k}}{\prod_{i=1}^k A_i^k} \lim_{n^* \to \infty} \frac{\Gamma\left(A_{k+1}^{n^*} + 1\right)}{\prod_{i=k+1}^{n^*} A_{k+1}^i}, \quad t \geq 0. \tag{5.13}$$

Here, empty sums are to be put to 0, empty products to 1. Our aim for the rest of this section is to show that (under appropriate conditions) this limit exists, and to prove that φ_0^∞ is a generator. This will involve a lot of algebraic details in order to justify the above iteration of limits. In particular, we need to impose some restrictions on the parametrizing sequence, so the following conditions for $\{\alpha_i\}_{i \in \mathbb{N}}$ are sufficient for the statements of this section to hold:

(R1) The sequence is increasing with $\alpha_1 > -1$ and convergent with limit α.

(R2) There exists $C \in \mathbb{R}$ such that

$$\sup_{n > k} \left| \ln(n-k) \sum_{i=1}^{n-k} \left(\frac{\alpha_{i+k}+1}{\alpha+1} - 1 \right) - \sum_{i=1}^{n-k} \frac{1}{i} \sum_{j=1}^i \left(\frac{\alpha_{j+k}+1}{\alpha+1} - 1 \right) \right| \;\leq\; C \tag{5.14}$$

for all $k \in \mathbb{N}_0$.

The convergence of $\{\alpha_i\}_{i \in \mathbb{N}}$ is a natural condition incorporated in the chosen approach since we deduce our results from those valid in the case of parametrizing sequences which are eventually constant and thus convergent. To require that $\{\alpha_i\}_{i \in \mathbb{N}}$ increases facilitates the proofs in the sequel but we believe that this condition might be omitted somehow. Condition (R2) concerns the speed of convergence of $\{\alpha_i\}_{i \in \mathbb{N}}$: For a series $\sum_{i=1}^\infty b_i$ with partial sums $S_1, S_2, \ldots$ consider the logarithmic means

$$\frac{1}{\ln n} \left(S_1 + \frac{S_2}{2} + \cdots \frac{S_n}{n} \right), \quad n \in \mathbb{N},$$

stemming from the theory of divergent series. Logarithmic means are regular, i.e. if the partial sums S_n converge, the logarithmic means converge versus the same limit, see for instance Hardy (1956). Thus with $b_i = \frac{\alpha_i+1}{\alpha+1} - 1$, $i \in \mathbb{N}$, we find

$$\sum_{i=1}^{n}\left(\frac{\alpha_i+1}{\alpha+1}-1\right) - \frac{1}{\ln n}\sum_{i=1}^{n}\frac{1}{i}\sum_{j=1}^{i}\left(\frac{\alpha_j+1}{\alpha+1}-1\right) \to 0 \qquad \text{for } n \to \infty \qquad (5.15)$$

which resembles the expressions under point (R2) for $k = 0$. Multiplication of (5.15) with the diverging term $\ln n$ makes clear that condition (R2) requires the logarithmic means to match the usual means sufficiently fast, that is, fast enough for the left side of (5.14) to be bounded.

Example 62: A sequence verifying conditions (R1) and (R2) is $\alpha_i = -q^i$, $i \in \mathbb{N}$, for some $0 \le q < 1$. Obviously, (R1) holds. To show (R2) we first prove that (5.14) is satisfied for $k = 0$. We have

$$-\sum_{i=1}^{n}\frac{1}{i}\sum_{j=1}^{i}\left(\frac{\alpha_j+1}{\alpha+1}-1\right) + \sum_{i=1}^{n}\left(\frac{\alpha_i+1}{\alpha+1}-1\right)\ln n \qquad (5.16)$$

$$= \sum_{i=1}^{n}\frac{1}{i}\sum_{j=1}^{i}q^j - \sum_{i=1}^{n}q^i\ln n$$

$$= \frac{q}{1-q}\left(\sum_{i=1}^{n}\frac{1-q^i}{i} - (1-q^n)\ln n\right), \qquad n \in \mathbb{N}.$$

Since $f(x) = \frac{1-q^x}{x}$ is decreasing for $x > 0$ because $f'(x) = \frac{q^x(1-\ln q^x)-1}{x^2} < 0$ for positive x we obtain due to Weierstraß's integral criterion for series

$$\int_{1}^{n+1}\frac{1-q^x}{x}dx - (1-q^n)\int_{1}^{n}\frac{1}{x}dx$$

$$\le \sum_{i=1}^{n}\frac{1-q^i}{i} - (1-q^n)\ln n$$

$$\le (1-q) + \int_{1}^{n}\frac{1-q^x}{x}dx - \int_{1}^{n}\frac{1}{x}dx + q^n\ln n$$

which roughly estimated implies

$$-\int_{1}^{n+1}q^x\,dx \le \sum_{i=1}^{n}\frac{1-q^i}{i} - (1-q^n)\ln n \le 1 + q^n\ln n, \qquad n \in \mathbb{N}.$$

Since $\int_{1}^{n+1}q^x\,dx$ as well as $1 + q^n\ln n$ are bounded by a constant, say $\tilde{C}$, for all $n \in \mathbb{N}$, we conclude that

$$\sum_{i=1}^{n}\frac{1}{i}\sum_{j=1}^{i}q^j - \sum_{i=1}^{n}q^i\ln n, \qquad n \in \mathbb{N},$$

and likewise expression (5.16) are bounded by $\frac{q}{1-q}\tilde{C}$. Further, for arbitrary $k \in \mathbb{N}_0$ the following holds:

$$\left| \sum_{i=1}^{n-k} \frac{1}{i} \sum_{j=1}^{i} \left(\frac{\alpha_{j+k}+1}{\alpha+1} - 1 \right) - \sum_{i=1}^{n-k} \left(\frac{\alpha_{i+k}+1}{\alpha+1} - 1 \right) \ln(n-k) \right|$$

$$= \left| \sum_{i=1}^{n-k} \frac{1}{i} \sum_{j=1}^{i} q^{j+k} - \sum_{i=1}^{n-k} q^{i+k} \ln(n-k) \right|$$

$$= q^k \left| \sum_{i=1}^{n-k} \frac{1}{i} \sum_{j=1}^{i} q^{j} - \sum_{i=1}^{n-k} q^{i} \ln(n-k) \right|, \qquad n > k,$$

which yields (R2) with $C = \frac{q}{1-q}\tilde{C}$ since $0 \le q < 1$. $\qquad\square$

Let us return to this section's aim which is to show that

$$\varphi_0^\infty(t \mid \alpha_1, \alpha_2, \ldots; \alpha)$$

$$= \lim_{n^* \to \infty} \frac{\prod_{i=1}^{n^*} A_1^i}{\Gamma(A_1^{n^*}+1)} \sum_{k=0}^{\infty} (-1)^k \frac{(\lambda t^{\alpha+1})^{A_1^k}}{\prod_{i=1}^{k} A_i^k} \lim_{n^* \to \infty} \frac{\Gamma(A_{k+1}^{n^*}+1)}{\prod_{i=k+1}^{n^*} A_{k+1}^i}, \qquad t \ge 0,$$

is a generator with respect to the sequence $\{\alpha_i\}_{i\in\mathbb{N}}$ subject to conditions (R1) and (R2). With regard to this main result we especially need to care for the expressions $\sum_{i=1}^{\infty} \left(\frac{\alpha_i+1}{\alpha+1} - 1 \right)$ and $\lim_{n^* \to \infty} \frac{\Gamma(A_{k+1}^{n^*}+1)}{\prod_{i=k+1}^{n^*} A_{k+1}^i}$. This is the issue of the following two lemmas:

Lemma 63: *Let $\{\alpha_i\}_{i\in\mathbb{N}}$ be a sequence such that conditions (R1) and (R2) hold. Then the limit*

$$A^\infty = \sum_{i=1}^{\infty} \left(\frac{\alpha_i+1}{\alpha+1} - 1 \right) = \frac{1}{\alpha+1} \sum_{i=1}^{\infty} (\alpha_i - \alpha) \tag{5.17}$$

exists.

Proof: W.l.o.g. $\alpha = 0$, else consider $\left\{ \frac{\alpha_i+1}{\alpha+1} - 1 \right\}_{i\in\mathbb{N}}$ instead of $\{\alpha_i\}_{i\in\mathbb{N}}$, which comes along with identical quantities $A_1^n - n$, $n \in \mathbb{N}$, and which satisfies (R1) and (R2) as well if $\{\alpha_i\}_{i\in\mathbb{N}}$ does.

Since $\{\alpha_i\}_{i\in\mathbb{N}}$ is increasing and converges to 0, we have $\alpha_i \le 0$, $i \in \mathbb{N}$. Therefore, $\sum_{i=1}^{n} \alpha_i$ decreases and $\sum_{i=1}^{n} \alpha_i \le 0$, $n \in \mathbb{N}$. To find a lower bound note firstly that for sufficiently large n we have

$$\sum_{i=1}^{n} \frac{1}{i} - \ln n \le \frac{3}{2}\gamma, \tag{5.18}$$

as $\gamma = \lim_{n\to\infty} \sum_{i=1}^{n} \frac{1}{i} - \ln n$ exists and defines Euler's constant which is positive, cp. Definition A.1. The above inequality (5.18) implies

$$\ln n - \sum_{j=i}^{n} \frac{1}{j} \;\geq\; \ln n - \sum_{j=3}^{n} \frac{1}{j} \;\geq\; \ln n - \int_{2}^{n} \frac{1}{x} dx \;=\; \ln 2, \qquad i \geq 3, \qquad (5.19)$$

which yields

$$\sum_{i=1}^{n} \alpha_i \;\geq\; \alpha_1 + \alpha_2 + \frac{1}{\ln 2} \sum_{i=3}^{n} \alpha_i \left(\ln n - \sum_{j=i}^{n} \frac{1}{j} \right)$$

$$= \; \alpha_1 \left(1 + \frac{\sum_{j=1}^{n} \frac{1}{j} - \ln n}{\ln 2} \right) + \alpha_2 \left(1 + \frac{\sum_{j=2}^{n} \frac{1}{j} - \ln n}{\ln 2} \right)$$

$$+ \frac{1}{\ln 2} \sum_{i=1}^{n} \alpha_i \left(\ln n - \sum_{j=i}^{n} \frac{1}{j} \right)$$

$$\geq \; \alpha_1 \left(1 + \frac{\frac{3}{2}\gamma}{\ln 2} \right) + \alpha_2 \left(1 + \frac{\frac{3}{2}\gamma - 1}{\ln 2} \right) - \frac{C}{\ln 2},$$

where the last inequality holds due to (5.18) and

$$\sum_{i=1}^{n} \alpha_i \left(\ln n - \sum_{j=i}^{n} \frac{1}{j} \right) \;=\; \ln n \sum_{i=1}^{n} \alpha_i - \sum_{i=1}^{n} \alpha_i \sum_{j=i}^{n} \frac{1}{j}$$

$$= \; \ln n \sum_{i=1}^{n} \alpha_i - \sum_{j=1}^{n} \frac{1}{j} \sum_{i=1}^{j} \alpha_i \;\overset{(R2)}{\geq}\; -C.$$

Thus $\sum_{i=1}^{n} \alpha_i$ is bounded for $n \in \mathbb{N}$ and due to its monotonicity convergent. $\blacksquare$

Lemma 64: *Let $\{\alpha_i\}_{i\in\mathbb{N}}$ be a sequence such that conditions (R1) and (R2) hold. Then for $k \in \mathbb{N}_0$ the limit*

$$\lim_{n^*\to\infty} \frac{\Gamma\left(A_{k+1}^{n^*} + 1\right)}{\prod_{i=k+1}^{n^*} A_{k+1}^{i}}$$

exists and further there exist $c_1, c_2 \in \mathbb{R}$ such that

$$0 < c_1 \leq \lim_{n^*\to\infty} \frac{\Gamma\left(A_{k+1}^{n^*} + 1\right)}{\prod_{i=k+1}^{n^*} A_{k+1}^{i}} \leq \sup_{n^*\in\mathbb{N}} \frac{\Gamma\left(A_{k+1}^{n^*} + 1\right)}{\prod_{i=k+1}^{n^*} A_{k+1}^{i}} \leq c_2$$

holds for all $k \in \mathbb{N}_0$.

Proof: Monotonicity: To begin with, notice that for $k \in \mathbb{N}_0$

$$\frac{\Gamma\left(A_{k+1}^{n^*} + 1\right)}{\prod_{i=k+1}^{n^*} A_{k+1}^{i}}$$

is decreasing in n^* for sufficiently large n^* since the quotient $\frac{\alpha_{n^*+1}+1}{\alpha+1}$ is at most 1 which yields

$$
\frac{\Gamma\left(A_{k+1}^{n^*+1}+1\right)}{\prod_{i=k+1}^{n^*+1} A_{k+1}^i} \cdot \frac{\prod_{i=k+1}^{n^*} A_{k+1}^i}{\Gamma\left(A_{k+1}^{n^*}+1\right)} = \frac{\Gamma\left(A_{k+1}^{n^*+1}+1\right)}{A_{k+1}^{n^*+1}\Gamma\left(A_{k+1}^{n^*}+1\right)}
$$

$$
= \frac{\Gamma\left(A_{k+1}^{n^*+1}\right)}{\Gamma\left(A_{k+1}^{n^*}+1\right)} = \frac{\Gamma\left(A_{k+1}^{n^*}+\frac{\alpha_{n^*+1}+1}{\alpha+1}\right)}{\Gamma\left(A_{k+1}^{n^*}+1\right)} \leq 1
$$

for n^* such that $A_{k+1}^{n^*+1}$ exceeds the extreme point of about 1.462 of the Gamma function in the first quadrant.

Upper bound: Let $l \in \mathbb{N}$ be the index such that $\frac{\alpha_l+1}{\alpha+1} \geq \frac{3}{4}$. Then we have for $n^* \geq l+1$

$$
\Gamma\left(A_{k+1}^{n^*}+1\right) = A_{k+1}^{n^*}\Gamma\left(A_{k+1}^{n^*}\right) = A_{k+1}^{n^*}\Gamma\left(A_{k+1}^{n^*-1}+\frac{\alpha_{n^*}+1}{\alpha+1}\right)
$$

$$
\leq A_{k+1}^{n^*}\Gamma\left(A_{k+1}^{n^*-1}+1\right), \qquad k < n^*-1,
$$

and by induction

$$
\Gamma\left(A_{k+1}^{n^*}+1\right) \leq A_{k+1}^{n^*}A_{k+1}^{n^*-1}\cdots A_{k+1}^{l+1}\Gamma\left(A_{k+1}^l+1\right), \qquad k < l,
$$

resp.

$$
\Gamma\left(A_{k+1}^{n^*}+1\right) \leq A_{k+1}^{n^*}A_{k+1}^{n^*-1}\cdots A_{k+1}^{k+2}\Gamma\left(A_{k+1}^{k+1}+1\right)
$$

$$
\leq A_{k+1}^{n^*}A_{k+1}^{n^*-1}\cdots A_{k+1}^{k+1}\Gamma\left(\frac{3}{4}\right), \qquad l \leq k < n^*-1.
$$

Altogether, we find

$$
\frac{\Gamma\left(A_{k+1}^{n^*}+1\right)}{\prod_{i=k+1}^{n^*} A_{k+1}^i} \leq \frac{\prod_{i=l+1}^{n^*} A_{k+1}^i \Gamma\left(A_{k+1}^l+1\right)}{\prod_{i=k+1}^{n^*} A_{k+1}^i} = \frac{\Gamma\left(A_{k+1}^l+1\right)}{\prod_{i=k+1}^{l} A_{k+1}^i}, \qquad k < l,
$$

resp.

$$
\frac{\Gamma\left(A_{k+1}^{n^*}+1\right)}{\prod_{i=k+1}^{n^*} A_{k+1}^i} \leq \Gamma\left(\frac{3}{4}\right), \qquad l \leq k < n^*-1.
$$

Hence $\frac{\Gamma\left(A_{k+1}^{n^*}+1\right)}{\prod_{i=k+1}^{n^*} A_{k+1}^i}$ is bounded uniformly for $k < n^*-1$ and $n^* \geq l+1$. Moreover, there also exists an upper bound for all $k \in \mathbb{N}_0$ and $n^* \in \mathbb{N}$ which exceeds additionally the finitely many values corresponding to

$$
(k,n^*) \in \{(i,j) \in \mathbb{N}_0 \times \mathbb{N} \,|\, l \leq j-1 \leq i \text{ or } j \leq l\}.
$$

Lower bound: First, notice that for $k \in \mathbb{N}_0$, $n^* \in \mathbb{N}$ with $n^* > k$

$$\frac{\Gamma\left(A_{k+1}^{n^*}+1\right)}{\prod_{i=k+1}^{n^*} A_{k+1}^i} = \frac{\Gamma\left(A_{k+1}^{n^*}+1\right)}{(n^*-k)! \prod_{i=k+1}^{n^*}\left(1+\frac{\sum_{j=k+1}^{i}\left(\frac{\alpha_j+1}{\alpha+1}-1\right)}{i-k}\right)}. \tag{5.20}$$

Stirling's formula, see equation (A.2), yields

$$\Gamma\left(A_{k+1}^{n^*}+1\right) = \sqrt{2\pi}\left(A_{k+1}^{n^*}\right)^{A_{k+1}^{n^*}+\frac{1}{2}} e^{-A_{k+1}^{n^*}+\frac{\vartheta_1}{12A_{k+1}^{n^*}}}$$

and

$$(n^*-k)! = \sqrt{2\pi}\,(n^*-k)^{n^*-k+\frac{1}{2}} e^{-(n^*-k)+\frac{\vartheta_2}{12(n^*-k)}}$$

for some $\vartheta_1, \vartheta_2 \in (0,1)$. Thus

$$\frac{\Gamma\left(A_{k+1}^{n^*}+1\right)}{(n^*-k)!}$$

$$\geq \left(A_{k+1}^{n^*}\right)^{A_{k+1}^{n^*}-(n^*-k)} \left(\frac{A_{k+1}^{n^*}}{n^*-k}\right)^{n^*-k+\frac{1}{2}} e^{-\left(A_{k+1}^{n^*}-(n^*-k)\right)-\frac{1}{12(n^*-k)}}$$

$$\geq \left(A_{k+1}^{n^*}\right)^{A_{k+1}^{n^*}-(n^*-k)} \sqrt{\frac{\alpha_1+1}{\alpha+1}} \left(\frac{A_{k+1}^{n^*}-(n^*-k)}{n^*-k}+1\right)^{n^*-k} e^{-\frac{1}{12(n^*-k)}}$$

since $\sqrt{\frac{A_{k+1}^{n^*}}{n^*-k}} \geq \sqrt{\frac{\alpha_1+1}{\alpha+1}}$ because $\{\alpha_i\}_{i\in\mathbb{N}}$ is increasing and since $e^{-\left(A_{k+1}^{n^*}-(n^*-k)\right)} \geq 1$ because $n^*-k \geq A_{k+1}^{n^*}$. For sufficiently large n^* (here we mean $n^* > k$ such that $\frac{A^\infty}{n^*-k} > -1$ with A^∞ given by (5.17)) we find

$$\frac{\Gamma\left(A_{k+1}^{n^*}+1\right)}{(n^*-k)!} \geq \left(A_{k+1}^{n^*}\right)^{A_{k+1}^{n^*}-(n^*-k)} \sqrt{\frac{\alpha_1+1}{\alpha+1}} \left(\frac{A^\infty}{n^*-k}+1\right)^{n^*-k} e^{-1}$$

$$\geq \tilde{c}\left(A_{k+1}^{n^*}\right)^{A_{k+1}^{n^*}-(n^*-k)}, \tag{5.21}$$

where the last inequality holds for some $\tilde{c} > 0$ independent of k since $\left(\frac{A^\infty}{n^*-k}+1\right)^{n^*-k}$ is increasing and approaches $\exp\{A^\infty\}$ for large n^*. Further, we find

$$\prod_{i=k+1}^{n^*}\left(1+\frac{\sum_{j=k+1}^{i}\left(\frac{\alpha_j+1}{\alpha+1}-1\right)}{i-k}\right) = \prod_{i=1}^{n^*-k}\left(1+\frac{\sum_{j=1}^{i}\left(\frac{\alpha_{j+k}+1}{\alpha+1}-1\right)}{i}\right)$$

$$= \exp\left\{\sum_{i=1}^{n^*-k} \ln\left(1+\frac{1}{i}\sum_{j=1}^{i}\left(\frac{\alpha_{j+k}+1}{\alpha+1}-1\right)\right)\right\}, \qquad k \in \mathbb{N}_0, n^* \in \mathbb{N}. \tag{5.22}$$

Since $\ln(1-x) \leq -x$ for $0 \leq x < 1$, we obtain

$$\exp\left\{\sum_{i=1}^{n^*-k} \ln\left(1 + \frac{1}{i}\sum_{j=1}^{i}\left(\frac{\alpha_{j+k}+1}{\alpha+1}-1\right)\right)\right\}$$

$$\leq \exp\left\{\sum_{i=1}^{n^*-k}\frac{1}{i}\sum_{j=1}^{i}\left(\frac{\alpha_{j+k}+1}{\alpha+1}-1\right)\right\}, \qquad (5.23)$$

for all $k \in \mathbb{N}_0$ and $n^* \in \mathbb{N}$. Combining equations (5.20)-(5.23) yields for $k \in \mathbb{N}_0$ and sufficiently large n^*

$$\frac{\Gamma\left(A_{k+1}^{n^*}+1\right)}{\prod_{i=k+1}^{n^*} A_{k+1}^i}$$

$$\geq \tilde{c}\cdot\exp\left\{\ln\left(A_{k+1}^{n^*}\right)\left(A_{k+1}^{n^*}-(n^*-k)\right) - \sum_{i=1}^{n^*-k}\frac{1}{i}\sum_{j=1}^{i}\left(\frac{\alpha_{j+k}+1}{\alpha+1}-1\right)\right\}.$$

With

$$\ln\left(A_{k+1}^{n^*}\right) = \underbrace{\ln\left(\frac{A_{k+1}^{n^*}-(n^*-k)}{n^*-k}+1\right)}_{\leq 0} + \ln\left(n^*-k\right) \leq \ln\left(n^*-k\right)$$

we finally obtain

$$\frac{\Gamma\left(A_{k+1}^{n^*}+1\right)}{\prod_{i=k+1}^{n^*} A_{k+1}^i}$$

$$\geq \tilde{c}\cdot\exp\left\{\ln\left(n^*-k\right)\sum_{i=1}^{n^*-k}\left(\frac{\alpha_{i+k}+1}{\alpha+1}-1\right) - \sum_{i=1}^{n^*-k}\frac{1}{i}\sum_{j=1}^{i}\left(\frac{\alpha_{j+k}+1}{\alpha+1}-1\right)\right\}$$

$$\overset{(R2)}{\geq} c_1,$$

for $k \in \mathbb{N}_0$, sufficiently large $n^* \in \mathbb{N}$ and some constant $c_1 > 0$ independent of k. Altogether for $k \in \mathbb{N}_0$ the limit

$$\lim_{n^*\to\infty}\frac{\Gamma\left(A_{k+1}^{n^*}+1\right)}{\prod_{i=k+1}^{n^*} A_{k+1}^i}$$

exists (since the sequence eventually decreases and at the same time is bounded below by e. g. 0) and with $c_1, c_2 > 0$ we find for all $k \in \mathbb{N}_0$ and $n^* \in \mathbb{N}$

$$0 < c_1 \leq \lim_{n^*\to\infty}\frac{\Gamma\left(A_{k+1}^{n^*}+1\right)}{\prod_{i=k+1}^{n^*} A_{k+1}^i} \leq \sup_{n^*\in\mathbb{N}}\frac{\Gamma\left(A_{k+1}^{n^*}+1\right)}{\prod_{i=k+1}^{n^*} A_{k+1}^i} \leq c_2. \qquad\blacksquare$$

Preparing the main result of the present section we deduce one more lemma to establish the uniform convergence of the series occurring in (5.12) resp. (5.13). It will justify especially the iteration of the limits in equation (5.13) subject to (R1) and (R2):

$$\varphi_0^\infty(t|\alpha_1, \alpha_2, \ldots; \alpha)$$

$$= \lim_{n^* \to \infty} \frac{\prod_{i=1}^{n^*} A_1^i}{\Gamma\left(A_1^{n^*} + 1\right)} \sum_{k=0}^\infty (-1)^k \frac{(\lambda t^{\alpha+1})^{A_1^k} \Gamma\left(A_{k+1}^{n^*} + 1\right)}{\prod_{i=1}^k A_i^k \prod_{i=k+1}^{n^*} A_{k+1}^i} \tag{5.24}$$

$$= \lim_{n^* \to \infty} \frac{\prod_{i=1}^{n^*} A_1^i}{\Gamma\left(A_1^{n^*} + 1\right)} \sum_{k=0}^\infty (-1)^k \frac{(\lambda t^{\alpha+1})^{A_1^k}}{\prod_{i=1}^k A_i^k} \lim_{n^* \to \infty} \frac{\Gamma\left(A_{k+1}^{n^*} + 1\right)}{\prod_{i=k+1}^{n^*} A_{k+1}^i}, \qquad t \geq 0. \tag{5.25}$$

Lemma 65: *Let $\{\alpha_i\}_{i \in \mathbb{N}}$ be a sequence such that (R1) and (R2) are satisfied. Then the following statements hold:*

a) Let $t \geq 0$. Then the series

$$\sum_{k=0}^\infty (-1)^k \frac{(\lambda t^{\alpha+1})^{A_1^k}}{\prod_{i=1}^k A_i^k} \cdot \frac{\Gamma\left(A_{k+1}^{n^*} + 1\right)}{\prod_{i=k+1}^{n^*} A_{k+1}^i} \tag{5.26}$$

converges uniformly for $n^* \in \mathbb{N}$.

b) Let $\bar{t} > 0$. The series

$$\sum_{k=0}^\infty (-1)^k \frac{(\lambda t^{\alpha+1})^{A_1^k}}{\prod_{i=1}^k A_i^k} \lim_{n^* \to \infty} \frac{\Gamma\left(A_{k+1}^{n^*} + 1\right)}{\prod_{i=k+1}^{n^*} A_{k+1}^i} \tag{5.27}$$

converges uniformly for $t \in [0, \bar{t}]$.

Proof: a) According to Weierstraß' criterion, for the uniform convergence of (5.26) for $n^* \in \mathbb{N}$ it suffices to show

$$\sum_{k=0}^\infty \sup_{n^* \in \mathbb{N}} \left| (-1)^k \frac{(\lambda t^{\alpha+1})^{A_1^k}}{\prod_{i=1}^k A_i^k} \cdot \frac{\Gamma\left(A_{k+1}^{n^*} + 1\right)}{\prod_{i=k+1}^{n^*} A_{k+1}^i} \right| < \infty, \qquad t \geq 0. \tag{5.28}$$

First, we find due to Lemma 64 that

$$\sum_{k=0}^\infty \sup_{n^* \in \mathbb{N}} \left| (-1)^k \frac{(\lambda t^{\alpha+1})^{A_1^k}}{\prod_{i=1}^k A_i^k} \cdot \frac{\Gamma\left(A_{k+1}^{n^*} + 1\right)}{\prod_{i=k+1}^{n^*} A_{k+1}^i} \right| \leq c \sum_{k=0}^\infty \frac{(\lambda t^{\alpha+1})^{A_1^k}}{\prod_{i=1}^k A_i^k}, \qquad t \geq 0,$$

for some constant $c > 0$. For the convergence of the series on the right hand side we study the quotient of successive summands, that is

$$\frac{(\lambda t^{\alpha+1})^{A_1^{k+1}}}{\prod_{i=1}^{k+1} A_i^{k+1}} \cdot \frac{\prod_{i=1}^k A_i^k}{(\lambda t^{\alpha+1})^{A_1^k}} = \frac{(\lambda t^{\alpha+1})^{\frac{\alpha_{k+1}+1}{\alpha+1}}}{\frac{\alpha_{k+1}+1}{\alpha+1} \prod_{i=1}^k \frac{A_i^{k+1}}{A_i^k}} = \frac{(\lambda t^{\alpha+1})^{\frac{\alpha_{k+1}+1}{\alpha+1}}}{\frac{\alpha_{k+1}+1}{\alpha+1} \prod_{i=1}^k \left(1 + \frac{\frac{\alpha_{k+1}+1}{\alpha+1}}{A_i^k}\right)}$$

$$\tag{5.29}$$

for $k \in \mathbb{N}_0$. Now $\frac{\alpha_i+1}{\alpha+1}$ tends to 1 for i tending to infinity and thus for arbitrary $\delta > 0$ there exists $I(\delta) \in \mathbb{N}$ such that $\left| \frac{\alpha_i+1}{\alpha+1} - 1 \right| < \delta$ for $i \geq I(\delta)$. This implies for sufficiently large $k \in \mathbb{N}_0$

$$\prod_{i=1}^{k} \left(1 + \frac{\frac{\alpha_{k+1}+1}{\alpha+1}}{A_i^k} \right) = \prod_{i=1}^{I(\delta)-1} \left(1 + \frac{\frac{\alpha_{k+1}+1}{\alpha+1}}{A_i^k} \right) \prod_{i=I(\delta)}^{k} \left(1 + \frac{\frac{\alpha_{k+1}+1}{\alpha+1}}{A_i^k} \right)$$

$$\geq \prod_{i=I(\delta)}^{k} \left(1 + \frac{1-\delta}{(k-i+1)(1+\delta)} \right),$$

which diverges when k tends to infinity. Hence, for k tending to ∞ expression (5.29) tends to 0 which proves (5.28) and finally a).

b) Subject to conditions (R1) and (R2) the series (5.27) converges uniformly for $t \in [0, \bar{t}]$ since

$$\sum_{k=0}^{\infty} \sup_{t \in [0,\bar{t}]} \left| (-1)^k \frac{(\lambda t^{\alpha+1})^{A_1^k}}{\prod_{i=1}^{k} A_i^k} \cdot \lim_{n^* \to \infty} \frac{\Gamma\left(A_{k+1}^{n^*}+1\right)}{\prod_{i=k+1}^{n^*} A_{k+1}^i} \right|$$

$$\leq \sum_{k=0}^{\infty} \frac{(\lambda \bar{t}^{\alpha+1})^{A_1^k}}{\prod_{i=1}^{k} A_i^k} \cdot \lim_{n^* \to \infty} \frac{\Gamma\left(A_{k+1}^{n^*}+1\right)}{\prod_{i=k+1}^{n^*} A_{k+1}^i} \overset{\text{Lemma 64}}{\leq} c \sum_{k=0}^{\infty} \frac{(\lambda \bar{t}^{\alpha+1})^{A_1^k}}{\prod_{i=1}^{k} A_i^k} < \infty,$$

where the series on the right side converges due to the considerations of part a). ∎

Finally, we are able to prove the following main result of the section:

Theorem 66: *Let $\{\alpha_i\}_{i\in\mathbb{N}}$ be such that conditions (R1) and (R2) hold. Then for $\lambda > 0$*

$$\varphi_0^{\infty}(t|\alpha_1,\alpha_2,\ldots;\alpha) = \beta_0^{-1} \sum_{k=0}^{\infty} (-1)^k \frac{(\lambda t^{\alpha+1})^{A_1^k}}{\prod_{i=1}^{k} A_i^k} \beta_k, \qquad t \geq 0,$$

where $\beta_k = \lim_{n^* \to \infty} \frac{\Gamma\left(A_{k+1}^{n^*}+1\right)}{\prod_{i=k+1}^{n^*} A_{k+1}^i}$, $k \in \mathbb{N}_0$, is a generator with respect to $\{\alpha_i\}_{i\in\mathbb{N}}$. The corresponding functions φ_n^{∞}, $n \in \mathbb{N}$, satisfy

$$\varphi_n^{\infty}(t|\alpha_1,\alpha_2,\ldots;\alpha) = \frac{\lambda^{A_1^n}(\alpha+1)^n}{\beta_0} \sum_{k=0}^{\infty} (-1)^k \frac{(\lambda t^{\alpha+1})^{A_{n+1}^{n+k}}}{\prod_{i=1}^{k} A_{n+i}^{n+k}} \beta_{n+k}, \quad t > 0. \quad (5.30)$$

Proof: To start with, let us recall the defining properties of a generator, cp. Definition 56: $\varphi_0^{\infty}(t|\alpha_1,\alpha_2,\ldots;\alpha)$, $t \geq 0$, is a generator w. r. t. the given sequence $\{\alpha_i\}_{i\in\mathbb{N}}$, if the family $\{\varphi_n^{\infty}\}_{n\in\mathbb{N}_0}$ recursively defined by

$$\varphi_{n+1}^{\infty}(t|\alpha_1,\alpha_2,\ldots;\alpha) = -t^{-\alpha_{n+1}}\dot{\varphi}_n^{\infty}(t|\alpha_1,\alpha_2,\ldots;\alpha), \qquad n \in \mathbb{N}_0, t > 0,$$

verifies

$\lim_{t\to 0} \varphi_0^\infty(t|\alpha_1, \alpha_2, \ldots; \alpha) = 1$,

$\varphi_n^\infty(t|\alpha_1, \alpha_2, \ldots; \alpha) \geq 0$ for $t > 0$ and

$\lim_{t\to\infty} \varphi_n^\infty(t|\alpha_1, \alpha_2, \ldots; \alpha) = 0$

for $n \in \mathbb{N}_0$. The first of the above three conditions can easily be seen: Since due to Lemma 65 b) the series $\sum_{k=0}^\infty (-1)^k \frac{(\lambda t^{\alpha+1})^{A_1^k}}{\prod_{i=1}^k A_i^k} \beta_k$ converges uniformly for e. g. $t \in [0,1]$, the function $\varphi_0^\infty(t|\alpha_1, \alpha_2, \ldots; \alpha)$ is right-continuous at $t = 0$, moreover

$$\varphi_0^\infty(0|\alpha_1, \alpha_2, \ldots; \alpha) = 1.$$

Next, let us show by induction that (5.30) holds for φ_n^∞, $n \in \mathbb{N}_0$. For $n = 0$ this is true by definition of φ_0^∞. Now let (5.30) hold for some $n \in \mathbb{N}_0$. Then

$$
\begin{aligned}
\varphi_{n+1}^\infty(t|\alpha_1, \alpha_2, \ldots; \alpha) &= -t^{-\alpha_{n+1}} \frac{d}{dt} \varphi_n^\infty(t|\alpha_1, \alpha_2, \ldots; \alpha) \\
&= -t^{-\alpha_{n+1}} \frac{d}{dt} \frac{\lambda^{A_1^n}(\alpha+1)^n}{\beta_0} \sum_{k=0}^\infty (-1)^k \frac{(\lambda t^{\alpha+1})^{A_{n+1}^{n+k}}}{\prod_{i=1}^k A_{n+i}^{n+k}} \beta_{n+k} \\
&= -t^{-\alpha_{n+1}} \frac{\lambda^{A_1^n}(\alpha+1)^n}{\beta_0} \sum_{k=0}^\infty \frac{d}{dt}(-1)^k \frac{(\lambda t^{\alpha+1})^{A_{n+1}^{n+k}}}{\prod_{i=1}^k A_{n+i}^{n+k}} \beta_{n+k} \qquad (5.31) \\
&= -t^{-\alpha_{n+1}} \frac{\lambda^{A_1^n}(\alpha+1)^n}{\beta_0} \sum_{k=1}^\infty (-1)^k \frac{\lambda(\alpha+1)t^\alpha A_{n+1}^{n+k}(\lambda t^{\alpha+1})^{A_{n+1}^{n+k}-1}}{\prod_{i=1}^k A_{n+i}^{n+k}} \beta_{n+k} \\
&= \frac{\lambda^{A_1^{n+1}}(\alpha+1)^{n+1}}{\beta_0} \sum_{k=1}^\infty (-1)^{k-1} \frac{(\lambda t^{\alpha+1})^{A_{n+2}^{n+k}}}{\prod_{i=2}^k A_{n+i}^{n+k}} \beta_{n+k} \\
&= \frac{\lambda^{A_1^{n+1}}(\alpha+1)^{n+1}}{\beta_0} \sum_{k=0}^\infty (-1)^k \frac{(\lambda t^{\alpha+1})^{A_{n+2}^{n+1+k}}}{\prod_{i=2}^{k+1} A_{n+i}^{n+1+k}} \beta_{n+1+k} \\
&= \frac{\lambda^{A_1^{n+1}}(\alpha+1)^{n+1}}{\beta_0} \sum_{k=0}^\infty (-1)^k \frac{(\lambda t^{\alpha+1})^{A_{n+1+1}^{n+1+k}}}{\prod_{i=1}^k A_{n+1+i}^{n+1+k}} \beta_{n+1+k}, \qquad t > 0.
\end{aligned}
$$

Interchanging differentiation and sum in line (5.31) is possible due to the uniform convergence of $\sum_{k=0}^\infty (-1)^k \frac{(\lambda t^{\alpha+1})^{A_{n+1}^{n+k}}}{\prod_{i=1}^k A_{n+i}^{n+k}} \beta_{n+k}$ and $\sum_{k=0}^\infty (-1)^k \frac{(\lambda t^{\alpha+1})^{A_{n+1+1}^{n+1+k}}}{\prod_{i=1}^k A_{n+1+i}^{n+1+k}} \beta_{n+1+k}$ on compact subsets of $[0,\infty)$ which is implied by Lemma 65 b) applied at $\{\alpha_{n+i}\}_{i\in\mathbb{N}}$ and $\{\alpha_{n+1+i}\}_{i\in\mathbb{N}}$. Hence (5.30) is satisfied for $n+1$ and thus holds for all $n \in \mathbb{N}_0$. As

$$\varphi_n^\infty(t|\alpha_1, \alpha_2, \ldots; \alpha) = \frac{\lambda^{A_1^n}(\alpha+1)^n \beta_n}{\beta_0} \varphi_0^\infty(t|\alpha_{n+1}, \alpha_{n+2}, \ldots; \alpha), \qquad t > 0, n \in \mathbb{N}_0, \tag{5.32}$$

it only remains to show for $n \in \mathbb{N}$ that

$$
\begin{aligned}
\varphi_0^\infty(t|\alpha_{n+1}, \alpha_{n+2}, \ldots; \alpha) &\geq 0, \qquad t > 0, \\
\lim_{t \to \infty} \varphi_0^\infty(t|\alpha_{n+1}, \alpha_{n+2}, \ldots; \alpha) &= 0,
\end{aligned}
\tag{5.33}
$$

which due to equation (5.32) imply the same to be true for $\varphi_n^\infty(\cdot|\alpha_1, \alpha_2, \ldots; \alpha)$. As the shifted sequences $\{\alpha_{n+i}\}_{i \in \mathbb{N}}$, $n \in \mathbb{N}$, satisfy likewise conditions (R1) and (R2) we restrict us to show (5.33) for $n = 0$:
The first condition is obviously verified since

$$
\varphi_0^\infty(t|\alpha_1, \alpha_2, \ldots; \alpha) = \lim_{n^* \to \infty} \varphi_0^{n^*}(t|\alpha_1, \ldots, \alpha_{n^*}; \alpha)
$$

and $\varphi_0^{n^*}(t|\alpha_1, \ldots, \alpha_{n^*}; \alpha) \geq 0$ for $n^* \in \mathbb{N}_0$, $t > 0$.

To prove $\lim_{t \to \infty} \varphi_0^\infty(t|\alpha_1, \alpha_2, \ldots; \alpha) = 0$ consider $\varphi_0^{n^*}(t|\alpha_1, \ldots, \alpha_{n^*}; \alpha)$ for arbitrary $n^* \in \mathbb{N}$ and recall that

$$
\varphi_0^{n^*}(t|\alpha_1, \ldots, \alpha_{n^*}; \alpha)
$$

$$
\overset{(5.11),(5.10)}{=} \frac{\lambda^{A_1^{n^*}} \prod_{i=1}^{n^*} \gamma_i}{\Gamma(A_1^{n^*} + 1)} \int_t^\infty \int_{s_1}^\infty \cdots \int_{s_{n^*-1}}^\infty \prod_{i=1}^{n^*} s_i^{\alpha_i} e^{-\lambda s_{n^*}^{\alpha+1}} \, ds_{n^*} ds_{n^*-1} \cdots ds_1
$$

$$
\overset{x_i = \lambda s_i^{\alpha+1}}{=} \frac{\prod_{i=1}^{n^*} A_1^i}{\Gamma(A_1^{n^*} + 1)} \int_{\lambda t^{\alpha+1}}^\infty \int_{x_1}^\infty \cdots \int_{x_{n^*-1}}^\infty \prod_{i=1}^{n^*} x_i^{\frac{\alpha_i+1}{\alpha+1} - 1} e^{-x_{n^*}} \, dx_{n^*} dx_{n^*-1} \cdots dx_1, \quad t > 0.
$$

Since $\int_t^\infty x^a e^{-x} dx \leq t^a e^{-t}$ for $t > 0$ and $a \leq 0$ we find

$$
\varphi_0^{n^*}(t|\alpha_1, \ldots, \alpha_{n^*}; \alpha) \leq \frac{\prod_{i=1}^{n^*} A_1^i}{\Gamma(A_1^{n^*} + 1)} \left(\lambda t^{\alpha+1}\right)^{A_1^{n^*} - n^*} e^{-\lambda t^{\alpha+1}}, \qquad t > 0, \, n^* \in \mathbb{N},
$$

which implies

$$
0 \leq \varphi_0^\infty(t|\alpha_1, \alpha_2, \ldots; \alpha) \leq \lim_{n^* \to \infty} \frac{\prod_{i=1}^{n^*} A_1^i}{\Gamma(A_1^{n^*} + 1)} \left(\lambda t^{\alpha+1}\right)^{A_1^{n^*} - n^*} e^{-\lambda t^{\alpha+1}}
$$

$$
= \frac{\left(\lambda t^{\alpha+1}\right)^{A^\infty}}{\beta_0} e^{-\lambda t^{\alpha+1}}, \qquad t > 0,
$$

where $\beta_0 > 0$, cp. Lemma 64. As the last expression tends to 0 when t tends to infinity so does φ_0^∞ and our proof is complete. ∎

In analogy to previous results, mixtures of the generator φ_0^∞ with respect to λ are generators as well:

Corollary 67: Let $\{\alpha_i\}_{i \in \mathbb{N}}$ be such that (R1) and (R2) hold and let W be a probability distribution on $(0, \infty)$. Then

$$\varphi_0(t) \;=\; \int_{(0,\infty)} \beta_0^{-1} \sum_{k=0}^{\infty} (-1)^k \frac{\left(\lambda t^{\alpha+1}\right)^{A_1^k}}{\prod_{i=1}^{k} A_i^k} \beta_k \, dW(\lambda), \qquad t \ge 0,$$

is a generator w. r. t. $\{\alpha_i\}_{i \in \mathbb{N}}$.

Proof: The result follows from Theorem 66 and Proposition C.6 saying that we can interchange differentiation and integration. ∎

We conclude this section with the remark that the obtained generators

$$\int_{(0,\infty)} \beta_0^{-1} \sum_{k=0}^{\infty} (-1)^k \frac{\left(\lambda t^{\alpha+1}\right)^{A_1^k}}{\prod_{i=1}^{k} A_i^k} \beta_k \, dW(\lambda), \qquad t \ge 0,$$

show similarities to the generator $\int_{(0,\infty)} e^{-\lambda t} dW(\lambda) = \int_{(0,\infty)} \sum_{k=0}^{\infty} (-1)^k \frac{(\lambda t)^k}{k!} dW(\lambda)$ which we found in the case of nonnegative parametrizing sequences. As already noted below Proposition 54, the terms A_1^k correspond to k and similarly $\prod_{i=1}^{k} A_i^k$ correspond to $k!$ which is a reasonable interpretation as $A_i^k - A_i^{k-1} = \frac{a_{k+1}}{a+1}$ approaches 1 for large k. The bounded terms β_k can be seen as some normalizing factors.

5.4   Generators with respect to periodic sequences

In the sequel we present a heuristic approach to deduce generators with respect to periodic sequences.

Consider a real sequence $\{\alpha_i\}_{i \in \mathbb{N}}$. Further, let N be a generalized order statistic process with respect to the given sequence and $\{\varphi_n\}_{n \in \mathbb{N}}$ such that

$$f_{T_1,\dots,T_n}(t_1,\dots,t_n) \;=\; \prod_{i=1}^{n} t_i^{\alpha_i} \cdot \varphi_n(t_n), \qquad n \in \mathbb{N}, \, 0 < t_1 \le \dots \le t_n.$$

The functions $\{\varphi_n\}_{n \in \mathbb{N}}$ are completely determined by the sequence $\{\alpha_i\}_{i \in \mathbb{N}}$ and the generator $\varphi_0(t) = P(T_1 > t | T_1 < \infty)$, which generally depends on the parameters. In the previous section we deduced the following generators w. r. t. parametrizing sequences satisfying conditions (R1) and (R2), i.e. which especially converge to some limit $\alpha > -1$:

$$\varphi_0(t) \;=\; \varphi_0^{\infty}(t | \alpha_1, \alpha_2, \dots ; \alpha) \;=\; \beta_0^{-1} \sum_{k=0}^{\infty} (-1)^k \frac{\left(\lambda t^{\alpha+1}\right)^{A_1^k}}{\prod_{i=1}^{k} A_i^k} \beta_k, \qquad t \ge 0,$$

cp. Theorem 66. The corresponding φ_n for $n \in \mathbb{N}$ satisfy

$$\varphi_n(t) \;=\; \varphi_n^{\infty}(t | \alpha_1, \alpha_2, \dots ; \alpha) \;=\; \frac{\lambda^{A_1^n} (\alpha+1)^n \beta_n}{\beta_0} \varphi_0^{\infty}(t | \alpha_{n+1}, \alpha_{n+2}, \dots ; \alpha), \quad t > 0,$$

compare (5.32).

Let us return to arbitrary parametrizing sequences. For a generator φ_0 w.r.t. a sequence $\{\alpha_i\}_{i \in \mathbb{N}}$ let us write alternatively $\varphi_0(\cdot | \alpha_1, \alpha_2, \ldots)$ to underline its dependence on the whole parametrizing sequence. The above representations of φ_n encourage us to presume that the functions φ_n, $n \in \mathbb{N}$, admit a structure similar to that of φ_0 differing only by a constant multiplicative factor and shifted parameters. That is

$$\varphi_n(t) = C_n \varphi_0(t | \alpha_{n+1}, \alpha_{n+2}, \ldots), \qquad t > 0,$$

where C_n denotes a positive constant. In the case of a periodic sequence $\{\alpha_i\}_{i \in \mathbb{N}}$, i.e. such that $\alpha_{i+p} = \alpha_i$ for $i \in \mathbb{N}$ and some period $p \in \mathbb{N}$, this approach implies especially that φ_{n+p} is proportional to φ_n since

$$\varphi_0(t | \alpha_{n+p+1}, \alpha_{n+p+2}, \ldots) = \varphi_0(t | \alpha_{n+1}, \alpha_{n+2}, \ldots), \qquad t > 0,\, n \in \mathbb{N}_0.$$

The recursion formula

$$\dot{\varphi}_n(t) = -\varphi_{n+1}(t)\, t^{\alpha_{n+1}}, \qquad t > 0,\, n \in \mathbb{N}_0,$$

yields

$$\dot{\varphi}_0(t | \alpha_1, \alpha_2, \ldots) = -t^{\alpha_1} C_1 \varphi_0(t | \alpha_2, \alpha_3, \ldots)$$

$$C_1 \dot{\varphi}_0(t | \alpha_2, \alpha_3, \ldots) = -t^{\alpha_2} C_2 \varphi_0(t | \alpha_3, \alpha_4, \ldots)$$

$$\vdots$$

$$C_{p-1} \dot{\varphi}_0(t | \alpha_p, \alpha_{p+1}, \ldots) = -t^{\alpha_p} C_p \varphi_0(t | \alpha_{p+1}, \alpha_{p+2}, \ldots) = -t^{\alpha_p} C_p \varphi_0(t | \alpha_1, \alpha_2, \ldots)$$

for $t > 0$ which can be represented in vectorial notation by

$$\ddot{\vec{\phi}}(t) = - \begin{pmatrix} 0 & C_1 t^{\alpha_1} & 0 & \cdots & 0 & 0 \\ 0 & 0 & \frac{C_2}{C_1} t^{\alpha_2} & \ddots & & 0 \\ \vdots & & 0 & \ddots & 0 & \vdots \\ \vdots & & & \ddots & \frac{C_{p-2}}{C_{p-3}} t^{\alpha_{p-2}} & 0 \\ 0 & 0 & 0 & \cdots & 0 & \frac{C_{p-1}}{C_{p-2}} t^{\alpha_{p-1}} \\ \frac{C_p}{C_{p-1}} t^{\alpha_p} & 0 & 0 & \cdots & 0 & 0 \end{pmatrix} \vec{\phi}(t), \qquad t > 0.$$

$$(5.34)$$

Thereby, $\vec{\phi}$ is a p-dimensional vector whose i-th component is $\varphi_0(\cdot | \alpha_i, \alpha_{i+1}, \ldots)$, $i = 1, \ldots, p$.

To find a generator with respect to arbitrary periodic sequences using the above approach means to search for a general solution of this first order system of linear differential equations subject to the conditions of integrability and nonnegativity of ϕ by components. However, we restrict us to discuss the following examples where the introduced approach is successfully applied in order to deduce generators with respect to two-periodic sequences:

Example 68: Consider the sequence $\{\alpha_i\}_{i\in\mathbb{N}}$ such that $\alpha_{2j-1} = 0$ and $\alpha_{2j} = 1$ for $j \in \mathbb{N}$. Then, system (5.34) has the form

$$\dot{\vec{\phi}} = -\begin{pmatrix} 0 & C_1 \\ \frac{C_2}{C_1}t & 0 \end{pmatrix} \vec{\phi}, \qquad t > 0,$$

and the following ordinary differential equations of second order hold for the components ϕ_1 and ϕ_2 of $\vec{\phi}$:

$$\ddot{\phi}_1 = C_2 t \phi_1, \tag{5.35}$$

$$\ddot{\phi}_2 = \frac{1}{t}\dot{\phi}_2 + C_2 t \phi_2, \qquad t > 0. \tag{5.36}$$

Equation (5.35) is solved by

$$\phi_1(t) = a\,\mathrm{Ai}\left(\sqrt[3]{C_2}t\right) + b\,\mathrm{Bi}\left(\sqrt[3]{C_2}t\right), \qquad t > 0,$$

for some constants $a, b \in \mathbb{C}$, where Ai and Bi denote the Airy functions, compare Appendix A.4. Since Ai as well as Bi are nonnegative on $[0,\infty)$ and further $\lim_{t\to\infty}\mathrm{Ai}(t) = 0$ and $\lim_{t\to\infty}\mathrm{Bi}(t) = \infty$, we finally find

$$\varphi_0(t) = \phi_1(t) = 3^{\frac{2}{3}}\Gamma\left(\frac{2}{3}\right)\mathrm{Ai}\left(\sqrt[3]{C_2}t\right), \qquad t \geq 0,$$

because φ_0 is required to be nonnegative with $\varphi_0(0) = 1$ and must approach 0 for large t. Put $C_2 = \lambda^3$ for some $\lambda > 0$ to obtain

$$\varphi_0(t) = 3^{\frac{2}{3}}\Gamma\left(\frac{2}{3}\right)\mathrm{Ai}(\lambda t) = \frac{\mathrm{Ai}(\lambda t)}{\mathrm{Ai}(0)}, \qquad t \geq 0.$$

Once we have determined $\varphi_0(t)$ it remains to examine if it actually is a generator with respect to the given sequence $\{\alpha_i\}_{i\in\mathbb{N}}$:

Denote by $\dot{\mathrm{Ai}}$ and $\ddot{\mathrm{Ai}}$ the first resp. second derivative of Ai and note that $-\dot{\mathrm{Ai}}(t) \geq 0$, $t \geq 0$, and $\lim_{t\to\infty}\dot{\mathrm{Ai}}(t) = 0$. For φ_1 we obtain

$$\varphi_1(t) = -\dot{\varphi}_0(t) = -\lambda\frac{\dot{\mathrm{Ai}}(\lambda t)}{\mathrm{Ai}(0)}, \qquad t > 0,$$

which is nonnegative and tends to 0 when t tends to infinity. For φ_2 we actually find

$$\varphi_2(t) = -\frac{\dot{\varphi}_1(t)}{t} = \frac{\lambda^2\ddot{\mathrm{Ai}}(\lambda t)}{\mathrm{Ai}(0)t} = \lambda^3\frac{\mathrm{Ai}(\lambda t)}{\mathrm{Ai}(0)} = \lambda^3\varphi_0(t), \qquad t > 0,$$

which like φ_0 is again nonnegative and approaches 0 for large t. Altogether we obtain

$$\varphi_{2j}(t) = \lambda^{3j}\varphi_0(t)$$

$$\varphi_{2j+1}(t) = \lambda^{3j}\varphi_1(t), \qquad j \in \mathbb{N}_0, t > 0,$$

so that conditions (5.3) and (5.4) hold for every φ_n, $n \in \mathbb{N}$. Consequently φ_0 is indeed a generator with respect to the given parametrizing sequence.

Let us check whether $\varphi_0(t)$ is a generator which can also be obtained as a Laplace transform (in the context of generators with respect to nonnegative sequences, compare Theorem 57): For the derivatives of φ_0 we find

$$\dot{\varphi}_0(t) = \frac{\lambda \dot{\text{Ai}}(\lambda t)}{\text{Ai}(0)},$$

$$\ddot{\varphi}_0(t) = \frac{\lambda^2 \ddot{\text{Ai}}(\lambda t)}{\text{Ai}(0)} = \frac{\lambda^3 t \text{Ai}(\lambda t)}{\text{Ai}(0)},$$

$$\varphi_0^{(3)}(t) = \frac{\lambda^3 \text{Ai}(\lambda t)}{\text{Ai}(0)} + \frac{\lambda^4 t \dot{\text{Ai}}(\lambda t)}{\text{Ai}(0)},$$

$$\varphi_0^{(4)}(t) = \frac{\lambda^4 \dot{\text{Ai}}(\lambda t)}{\text{Ai}(0)} + \frac{\lambda^5 t \ddot{\text{Ai}}(\lambda t) + \lambda^4 \dot{\text{Ai}}(\lambda t)}{\text{Ai}(0)} = \frac{\lambda^6 t^2 \text{Ai}(\lambda t)}{\text{Ai}(0)} + \frac{2\lambda^4 \dot{\text{Ai}}(\lambda t)}{\text{Ai}(0)}, \quad t > 0,$$

where the fourth derivative changes its sign between $t = \frac{1}{\lambda}$ and $t = \frac{2}{\lambda}$. Hence φ_0 cannot be obtained as a Laplace transform which had to be completely monotone.

Thanks to Proposition C.6, which enables us to interchange differentiation and integration, along with φ_0 we obtain further generators as mixtures w. r. t. λ, that is

$$\int_{(0,\infty)} \frac{\text{Ai}(\lambda t)}{\text{Ai}(0)} dW(\lambda), \quad t \geq 0,$$

where W is some probability measure concentrated on $(0, \infty)$, is again a generator w.r.t the given parametrizing sequence. $\qquad\square$

Especially in the light of Section 4.3's nonexistence example (referring to a parametrizing sequence eventually constant -1), the next example is of great interest. Aiming at the question, whether or not we can find a GOS-process w. r. t. a sequence consisting of an infinite number of elements less or equal to -1, we study a 2-periodic parametrizing sequence which alternates between the elements 0 and -1. Our approach turns out to be successful:

Example 69: Consider the sequence $\{\alpha_i\}_{i\in\mathbb{N}}$ such that $\alpha_{2j-1} = 0$ and $\alpha_{2j} = -1$ for $j \in \mathbb{N}$. This time (5.34) has the form

$$\ddot{\vec{\phi}} = -\begin{pmatrix} 0 & C_1 \\ \frac{C_2}{C_1 t} & 0 \end{pmatrix} \vec{\phi}, \quad t > 0,$$

and the following ordinary differential equations of second order hold:

$$\ddot{\phi}_1 = \frac{C_2}{t} \phi_1, \qquad \ddot{\phi}_2 = -\frac{1}{t}\dot{\phi}_2 + \frac{C_2}{t}\phi_2, \quad t > 0.$$

The transformed function $\tilde{\phi}_1(t) = t^{-1}\phi_1\left(\left(\frac{t}{2}\right)^2/C_2\right)$ satisfies the characteristic differential equation (A.6) defining Bessel functions, and thus finally leads to the following solution for ϕ_1

$$\phi_1(t) \;=\; a\sqrt{C_2 t}\,\mathrm{BesselI}_1\left(2\sqrt{C_2 t}\right) + b\sqrt{C_2 t}\,\mathrm{BesselK}_1\left(2\sqrt{C_2 t}\right), \qquad t > 0,$$

for some constants $a, b \in \mathbb{C}$, where BesselI resp. BesselK denote the modified Bessel functions of the first resp. second kind, compare Appendix A.5. Since $\mathrm{BesselK}_1$ as well as $\mathrm{BesselI}_1$ are nonnegative on $(0, \infty)$, further $\lim_{t\to\infty}\mathrm{BesselK}_1(t) = 0$ and $\lim_{t\to\infty}\mathrm{BesselI}_1(t) = \infty$, we have

$$\varphi_0(t) \;=\; \phi_1(t) \;=\; 2\sqrt{C_2 t}\,\mathrm{BesselK}_1\left(2\sqrt{C_2 t}\right), \qquad t > 0, \tag{5.37}$$

because φ_0 is required to be nonnegative with $\lim_{t\to 0}\varphi_0(t) = 1$ and $\lim_{t\to\infty}\varphi_0(t) = 0$. For φ_1 we then obtain with $\lambda = C_2$

$$\varphi_1(t) \;=\; -\dot{\varphi}_0(t) \;=\; -\frac{d}{dt}2\sqrt{\lambda t}\,\mathrm{BesselK}_1\left(2\sqrt{\lambda t}\right)$$

$$\overset{(A.7)}{=}\; \frac{d}{dt}2\sqrt{\lambda t}\,\dot{\mathrm{BesselK}}_0\left(2\sqrt{\lambda t}\right)$$

$$=\; \sqrt{\frac{\lambda}{t}}\,\dot{\mathrm{BesselK}}_0\left(2\sqrt{\lambda t}\right) + 2\sqrt{\lambda t}\,\ddot{\mathrm{BesselK}}_0\left(2\sqrt{\lambda t}\right)\cdot\sqrt{\frac{\lambda}{t}}$$

$$=\; \frac{1}{2t}\left(2\sqrt{\lambda t}\,\dot{\mathrm{BesselK}}_0\left(2\sqrt{\lambda t}\right) + \left(2\sqrt{\lambda t}\right)^2\ddot{\mathrm{BesselK}}_0\left(2\sqrt{\lambda t}\right)\right)$$

$$\overset{(A.6)}{=}\; 2\lambda\,\mathrm{BesselK}_0\left(2\sqrt{\lambda t}\right), \qquad t > 0,$$

which is nonnegative and tends to 0 when t tends to infinity. For φ_2 we actually have

$$\varphi_2(t) \;=\; -\dot{\varphi}_1(t)t \;=\; -t\frac{d}{dt}2\lambda\,\mathrm{BesselK}_0\left(2\sqrt{\lambda t}\right)$$

$$\overset{(A.7)}{=}\; t2\lambda\,\mathrm{BesselK}_1\left(2\sqrt{\lambda t}\right)\sqrt{\frac{\lambda}{t}}$$

$$=\; \lambda\cdot 2\sqrt{\lambda t}\,\mathrm{BesselK}_1\left(2\sqrt{\lambda t}\right) \;=\; \lambda\varphi_0(t), \qquad t > 0.$$

Altogether, φ_0 as given by (5.37) is a generator w. r. t. the 2-periodic sequence alternating between 0 and -1. The corresponding functions φ_n, $n \in \mathbb{N}_0$, are given by

$$\varphi_{2j}(t) \;=\; \lambda^j\varphi_0(t) \;=\; \lambda^j 2\sqrt{\lambda t}\,\mathrm{BesselK}_1\left(2\sqrt{\lambda t}\right),$$

$$\varphi_{2j+1}(t) \;=\; \lambda^j\varphi_1(t) \;=\; 2\lambda^{j+1}\mathrm{BesselK}_0\left(2\sqrt{\lambda t}\right), \qquad j \in \mathbb{N}_0, t > 0.$$

Again, along with φ_0 we obtain the following generators as mixtures w.r.t. λ and w.r.t. arbitrary probability measures W on $(0, \infty)$ due to Proposition C.6:

$$\int_{(0,\infty)} 2\sqrt{\lambda t}\, \text{BesselK}_1\left(2\sqrt{\lambda t}\right) dW(\lambda), \qquad t > 0.$$

$\square$

To conclude with, remark that the studied 2-periodic parametrizing sequences of Examples 68 and 69 can also be seen as $2m$-periodic sequences for arbitrary $m \in \mathbb{N}$. One may wonder if the generators obtained by the presented approach for a p-periodic parametrizing sequence differ from those obtained likewise if the sequence is interpreted as $(m \cdot p)$-periodic sequence for some $m \in \mathbb{N}$. Further, one may ask how these possibly different generators are connected. The next section attempts to answer this last question.

5.5   A martingale characterization of generators

The last subsections were essentially dedicated to present generators in various special cases. As we already noted, given a certain sequence, the corresponding generator is not necessarily unique. So, for instance with respect to the sequence alternating between 0 and 1 we found the generators

$$\varphi_0(t) \;=\; \int_{(0,\infty)} e^{-\lambda t} dW(\lambda), \qquad t \geq 0,$$

for arbitrary distributions W on $(0, \infty)$ in the context of generators with respect to nonnegative parametrizing sequences, compare Section 5.1, as well as generators of the form

$$\varphi_0(t) \;=\; \int_{(0,\infty)} \frac{\text{Ai}(\lambda t)}{\text{Ai}(0)} dW(\lambda), \qquad t \geq 0,$$

for arbitrary W, obtained in the context of generators with respect to periodic parametrizing sequences, compare the previous section.

The present section highlights the question how different generators with respect to one and the same sequence are connected.

For simplification, in this section we consider a canonical model of point processes such that $(\Omega, \mathcal{F}) = (\mathcal{N}, \mathcal{H}(\mathcal{N}))$ is endowed with the family $\mathcal{P}$ of all probability measures on $(\Omega, \mathcal{F})$ and where N is the canonical point process, i.e. $N_t(\omega) = \omega(t)$ for $t \geq 0$ and $\omega \in \Omega$. Analogously we denote by $T_1, T_2, \ldots$ the occurrence times of N. Let further $\{\mathcal{F}_t\}_{t \geq 0}$ be the natural filtration of N. For a measure $P \in \mathcal{P}$ and for $t \geq 0$ we denote by P^t the restriction of P on $\mathcal{F}_t$.

In this context the canonical process N is just a formal component adequate for our distribution based considerations. Information about the distribution is carried completely by the measures $P \in \mathcal{P}$.

Denote by $\mathcal{G}$ the subset of $\mathcal{P}$ of measures such that N satisfies the generalized order statistic property, further, almost surely does not explode and such that the event $\{T_1 < \infty\}$ has probability 1. That is, the studies of this section will be restricted to point processes which are finite on finite intervals and which jump almost surely. In particular, according to Remark 37 the functions $\{\varphi_n\}_{n\in\mathbb{N}_0}$ corresponding to a nonexploding GOS-process are positive. Note further that to assume the event $\{T_1 = \infty\}$ to be a null set is sufficient for considerations concerning generators as these naturally correspond to GOS-processes which jump almost surely.

We find that for two measures $P, Q \in \mathcal{G}$ the restrictions P^t and Q^t to $\mathcal{F}_t$ are equivalent for each $t \geq 0$:

Proposition 70: *Let $P, Q \in \mathcal{G}$ be two generalized order statistic processes w. r. t. the sequences $\{\alpha_i^P\}_{i\in\mathbb{N}}$ resp. $\{\alpha_i^Q\}_{i\in\mathbb{N}}$ and with corresponding functions $\{\varphi_n^P\}_{n\in\mathbb{N}}$ resp. $\{\varphi_n^Q\}_{n\in\mathbb{N}}$. Then P^t is absolutely continuous with respect to Q^t and the corresponding Radon-Nikodým-densities verify*

$$\frac{dQ^t}{dP^t} = \prod_{i=1}^{N_t} T_i^{\alpha_i^Q - \alpha_i^P} \frac{\varphi_{N_t}^Q(t)}{\varphi_{N_t}^P(t)} \qquad P^t\text{-a. s.}$$

for $t \geq 0$.

Proof: For $n \in \mathbb{N}_0$ and $0 \leq t_1 \leq \cdots \leq t_n \leq t$ define the sets A, A_j and B_j, $j \in \mathbb{N}_0$ as follows:

$$A = \{\omega \in \Omega \,|\, T_1(\omega) \leq t_1, \ldots, T_n(\omega) \leq t_n\} \in \mathcal{F}_t,$$

$$A_j = A \cap \{\omega \in \Omega \,|\, \omega(t) = j\} \in \mathcal{F}_t,$$

$$B_j = \{(s_1, \ldots, s_{j+1}) \in [0, \infty)^{j+1} \,|\, 0 \leq s_1 \leq t_1, s_{i-1} < s_i \leq t_i, i = 2, \ldots, j, t < s_{j+1}\}.$$

Then we find

$$\int_A \prod_{i=1}^{N_t} T_i^{\alpha_i^Q - \alpha_i^P} \frac{\varphi_{N_t}^Q(t)}{\varphi_{N_t}^P(t)} \, dP^t = \sum_{j=n}^{\infty} \int_{A_j} \prod_{i=1}^{j} T_i^{\alpha_i^Q - \alpha_i^P} \frac{\varphi_j^Q(t)}{\varphi_j^P(t)} \, dP$$

$$= \sum_{j=n}^{\infty} \int_{\Omega} \prod_{i=1}^{j} T_i^{\alpha_i^Q - \alpha_i^P} \frac{\varphi_j^Q(t)}{\varphi_j^P(t)} \mathbb{1}_{B_j}(T_1, \ldots, T_{j+1}) \, dP$$

$$= \sum_{j=n}^{\infty} \int_{B_j} \prod_{i=1}^{j} s_i^{\alpha_i^Q - \alpha_i^P} \frac{\varphi_j^Q(t)}{\varphi_j^P(t)} \, dP_{T_1, \ldots, T_{j+1}}(s_1, \ldots, s_{j+1})$$

$$\stackrel{(3.7)}{=} \sum_{j=n}^{\infty} \int_{B_j} \prod_{i=1}^{j} s_i^{\alpha_i^Q - \alpha_i^P} \prod_{i=1}^{j+1} s_i^{\alpha_i^P} \varphi_{j+1}^P(s_{j+1}) \, d(s_1, \ldots, s_{j+1}) \frac{\varphi_j^Q(t)}{\varphi_j^P(t)}$$

$$= \sum_{j=n}^{\infty} \int_0^{t_1} \int_{s_1}^{t_2} \cdots \int_{s_{j-1}}^{t_j} \int_t^{\infty} \prod_{i=1}^{j} s_i^{\alpha_i^Q} s_{j+1}^{\alpha_{j+1}^P} \varphi_{j+1}^P(s_{j+1}) \, ds_{j+1} \cdots ds_1 \frac{\varphi_j^Q(t)}{\varphi_j^P(t)}$$

$$\overset{(3.10)}{=} \sum_{j=n}^{\infty} \int_0^{t_1} \int_{s_1}^{t_2} \cdots \int_{s_{j-1}}^{t_j} \prod_{i=1}^{j} s_i^{\alpha_i^Q} \, ds_j \cdots ds_1 \varphi_j^Q(t)$$

$$\overset{(3.10)}{=} \sum_{j=n}^{\infty} \int_0^{t_1} \int_{s_1}^{t_2} \cdots \int_{s_{j-1}}^{t_j} \int_t^{\infty} \prod_{i=1}^{j+1} s_i^{\alpha_i^Q} \varphi_{j+1}^Q(s_{j+1}) \, ds_{j+1} \cdots ds_1$$

$$= \sum_{j=n}^{\infty} Q(A_j) = Q^t(A).$$

Now for $t \geq 0$ let $\mathcal{Z}_t$ be the family of sets with a structure similar to A, that is

$$\mathcal{Z}_t = \big\{ \{\omega \in \Omega \,|\, T_1(\omega) \leq t_1, \ldots, T_n(\omega) \leq t_n\}, \, n \in \mathbb{N}_0, \, 0 \leq t_1 \leq \ldots \leq t_n \leq t \big\}.$$

Then $\mathcal{Z}_t$ contains Ω, is closed with respect to intersections and generates $\mathcal{F}_t$, $t \geq 0$. Therefore we have

$$Q^t(A) = \int_A \prod_{i=1}^{N_t} T_i^{\alpha_i^Q - \alpha_i^P} \frac{\varphi_{N_t}^Q(t)}{\varphi_{N_t}^P(t)} \, dP^t, \qquad A \in \mathcal{Z}_t, \, t \geq 0,$$

which, due to the uniqueness theorem of measures, implies

$$Q^t(A) = \int_A \prod_{i=1}^{N_t} T_i^{\alpha_i^Q - \alpha_i^P} \frac{\varphi_{N_t}^Q(t)}{\varphi_{N_t}^P(t)} \, dP^t, \qquad A \in \mathcal{F}_t, \, t \geq 0,$$

and the proof is complete. ∎

Corollary 71: *Let $P, Q \in \mathcal{G}$. Then under P the process*

$$\left\{ \prod_{i=1}^{N_t} T_i^{\alpha_i^Q - \alpha_i^P} \frac{\varphi_{N_t}^Q(t)}{\varphi_{N_t}^P(t)} \right\}_{t \geq 0}$$

forms an $\{\mathcal{F}_t\}_{t \geq 0}$-martingale.

Proof: Let $t \geq 0$. Further consider $s \in [0, t]$ and a set $A \in \mathcal{F}_s$. As the given process is a process of Radon-Nikodým densities, i. e.

$$\left\{ \frac{dQ^t}{dP^t} \right\}_{t \geq 0} = \left\{ \prod_{i=1}^{N_t} T_i^{\alpha_i^Q - \alpha_i^P} \frac{\varphi_{N_t}^Q(t)}{\varphi_{N_t}^P(t)} \right\}_{t \geq 0},$$

we obtain

$$\int_A \frac{dQ^t}{dP^t} \, dP = \int_A \frac{dQ^t}{dP^t} \, dP^t = \int_A dQ^t = \int_A dQ^s = \int_A \frac{dQ^s}{dP^s} \, dP^s = \int_A \frac{dQ^s}{dP^s} \, dP.$$

As a Radon-Nikodým density, $\prod_{i=1}^{N_t} T_i^{\alpha_i^Q - \alpha_i^P} \frac{\varphi_{N_t}^Q(t)}{\varphi_{N_t}^P(t)}$ is $\mathcal{F}_t$-measurable and integrable for all $t \geq 0$. ∎

Especially handsome martingales turn out if the sequences $\{\alpha_i^P\}_{i\in\mathbb{N}}$ and $\{\alpha_i^Q\}_{i\in\mathbb{N}}$ coincide. Then we find that for two different generators φ_0^P and φ_0^Q the process

$$\left\{ \frac{\varphi_{N_t}^Q(t)}{\varphi_{N_t}^P(t)} \right\}_{t\geq 0}$$

is an $\{\mathcal{F}_t\}_{t\geq 0}$-martingale. This property is even characteristic:

Theorem 72: *Let $P, Q \in \mathcal{G}$ be two generalized order statistic processes with respect to $\{\alpha_i^P\}_{i\in\mathbb{N}}$ resp. $\{\alpha_i^Q\}_{i\in\mathbb{N}}$ and with corresponding functions $\{\varphi_n^P\}_{n\in\mathbb{N}}$ resp. $\{\varphi_n^Q\}_{n\in\mathbb{N}}$. Then the following statements are equivalent:*

(i) The two sequences coincide, i. e. $\alpha_i^P = \alpha_i^Q$ for $i \in \mathbb{N}$.

(ii) Under P, the process

$$\left\{ \frac{\varphi_{N_t}^Q(t)}{\varphi_{N_t}^P(t)} \right\}_{t\geq 0} \tag{5.38}$$

is a martingale with respect to $\{\mathcal{F}_t\}_{t\geq 0}$.

Proof: With Corollary 71 it only remains to show that (ii) implies (i):

Let $\left\{ \dfrac{\varphi_{N_t}^Q(t)}{\varphi_{N_t}^P(t)} \right\}_{t\geq 0}$ be an $\{\mathcal{F}_t\}_{t\geq 0}$-martingale. Then

$$\sum_{k=j}^{\infty} \frac{\varphi_k^Q(t)}{\varphi_k^P(t)} P(N_t = k | N_s = j) = \frac{\varphi_j^Q(s)}{\varphi_j^P(s)} \quad \text{for } j \in \mathbb{N}_0 \text{ and } 0 < s \leq t.$$

The conditional probabilities equal

$$P(N_t = k | N_s = j) \overset{(3.22)}{=} \begin{cases} \Psi_{k-j}\left(s, t | \alpha_{j+1}^P, \ldots, \alpha_k^P\right) \dfrac{\varphi_k^P(t)}{\varphi_j^P(s)} & \text{if } k \geq j, \\ 0 & \text{else,} \end{cases}$$

for $j, k \in \mathbb{N}_0$ and $0 < s \leq t$ and where $\Psi_0(x,y) = 1$ respectively

$$\Psi_k(x, y | \alpha_1, \ldots, \alpha_k) = \int_x^y \int_{s_1}^y \cdots \int_{s_{k-1}}^y \prod_{j=1}^k s_j^{\alpha_j} ds_k \cdots ds_1, \qquad 0 < x \leq y.$$

Hence, we find

$$\sum_{k=j}^{\infty} \frac{\varphi_k^Q(t)}{\varphi_k^P(t)} \Psi_{k-j}\left(s, t | \alpha_{j+1}^P, \ldots, \alpha_k^P\right) \frac{\varphi_k^P(t)}{\varphi_j^P(s)} = \frac{\varphi_j^Q(s)}{\varphi_j^P(s)}$$

or equivalently

$$\sum_{k=j}^{\infty} \varphi_k^Q(t) \Psi_{k-j}\left(s, t | \alpha_{j+1}^P, \ldots, \alpha_k^P\right) = \varphi_j^Q(s), \qquad j \in \mathbb{N}_0, 0 < s \leq t. \tag{5.39}$$

Note that for fixed $t > 0$ the above series converges uniformly for $s \in [b,t]$, $b > 0$, due to Weierstraß' criterion for the uniform convergence of series as

$$\sum_{k=j}^{\infty} \sup_{s \in [b,t]} \left| \varphi_k^Q(t) \Psi_{k-j} \left(s,t \middle| \alpha_{j+1}^P, \ldots, \alpha_k^P \right) \right|$$

$$= \sum_{k=j}^{\infty} \varphi_k^Q(t) \Psi_{k-j} \left(b,t \middle| \alpha_{j+1}^P, \ldots, \alpha_k^P \right) \overset{(5.39)}{=} \varphi_j^Q(b) < \infty$$

holds since $\Psi_{k-j}\left(s,t \middle| \alpha_{j+1}^P, \ldots, \alpha_k^P \right)$ is decreasing in s. To prove (i) we differentiate equation (5.39) with respect to s and obtain for $j \in \mathbb{N}_0$ and $0 < s \leq t$

$$- \sum_{k=j+1}^{\infty} \varphi_k^Q(t) \Psi_{k-j-1} \left(s,t \middle| \alpha_{j+2}^P, \ldots, \alpha_k^P \right) \cdot s^{\alpha_{j+1}^P} = -s^{\alpha_{j+1}^Q} \varphi_{j+1}^Q(s) \qquad (5.40)$$

due to (3.11) and $\frac{d}{ds}\Psi_n(s,t|\alpha_1, \ldots, \alpha_n) = -\Psi_{n-1}(s,t|\alpha_2, \ldots, \alpha_n)s^{\alpha_1}$ for $n \in \mathbb{N}$. Thereby summation and differentiation can be interchanged since the above series converges uniformly for $s \in [b,t]$, $b > 0$, as it equals (5.39) with $j+1$ instead of j and multiplied with $s^{\alpha_{j+1}^P}$. Combining equation (5.39) with $j+1$ instead of j and (5.40) yields

$$s^{\alpha_{j+1}^P} \varphi_{j+1}^Q(s) = s^{\alpha_{j+1}^Q} \varphi_{j+1}^Q(s), \qquad 0 < s \leq t,$$

which is true if and only if $\alpha_{j+1}^P = \alpha_{j+1}^Q$, since φ_{j+1}^Q is positive on $(0,\infty)$. As j is an arbitrary number in $\mathbb{N}_0$ the proof is complete. ∎

We conclude the chapter by an application of Theorem 72 to mixed Poisson processes:

Some martingales connected with mixed Poisson processes

For mixed Poisson processes which are GOS-processes with respect to the sequence $\alpha_i = 0$ for $i \in \mathbb{N}$, every generator can be represented by

$$\varphi_0(t) = \int_{(0,\infty)} e^{-\lambda t} \, dW(\lambda), \qquad t \geq 0,$$

where W is some distribution on $(0,\infty)$, further

$$\varphi_n(t) = \int_{(0,\infty)} \lambda^n e^{-\lambda t} \, dW(\lambda), \qquad n \in \mathbb{N}_0, \, t > 0,$$

compare Theorem 40. Given two different distributions W_P and W_Q, by Theorem 72 we find the martingale

$$\left\{ \frac{\int_{(0,\infty)} \lambda^{N_t} e^{-\lambda t} dW_Q(\lambda)}{\int_{(0,\infty)} \lambda^{N_t} e^{-\lambda t} dW_P(\lambda)} \right\}_{t \geq 0} \qquad (5.41)$$

provided that N is a mixed Poisson process with mixing distribution W_P.

Remark that in Niese (2006) we can find a characterization of those classes of mixed Poisson processes, whose Radon-Nikodým densities given in (5.41) admit an exponential structure.

5.6   Existence of generalized order statistic processes – a summary

As the construction of generators is directly linked to the question whether generalized order statistic processes w. r. t. a given parametrizing sequence exist or not (compare the discussion following Definition 56), we end the chapter with a short survey of the knowledge gained until now concerning the existence of GOS-processes:

Beginning in Chapter 4 we showed that GOS-processes w. r. t. certain nonstandard parametrizing sequences exist by fully specifying and characterizing their distribution. So, Proposition 39 and Theorem 40 implicitly state the existence of GOS-processes w. r. t. eventually constant sequences with constants exceeding -1. The remaining case of parametrizing sequences eventually constant -1 was treated in Proposition 44 showing that corresponding (nontrivial) GOS-processes cannot exist.

In general, it remains an open question w. r. t. which parametrizing sequences there do or do not exist GOS-processes. Apparently, parameters α_i that equal -1 seem to be perturbing. However, knowing that it is not only an infinite number of parameters equal to -1 which corrupts the existence, cp. Example 69, we can only suggest that e. g. the boundedness of the corresponding sequence $\{\gamma_i\}_{i \in \mathbb{N}}$ or some similar condition might mark the border between existence and nonexistence of corresponding GOS-processes.

Nevertheless, beyond the case of eventually constant parametrizing sequences, we were able to implicitly prove the existence of GOS-processes in three special cases by constructing generators based on which GOS-processes can be obtained:

So, w. r. t. nonnegative parametrizing sequences a whole class of generators is given by Laplace transforms of probability measures W on $(0, \infty)$,

$$\varphi_0(t) = \int_{(0,\infty)} e^{-\lambda t} dW(\lambda), \qquad t \geq 0,$$

compare Theorem 57. However there are generators which cannot be represented likewise, i. e., there are GOS-processes w. r. t. nonnegative sequences whose distributions are not generated by Laplace transforms.

With respect to increasing, convergent parametrizing sequences satisfying condition (R2) of Section 5.3 and with the help of limit considerations, compare Theorem 66 resp. Corollary 67, we were able to deduce generators of the form

$$\varphi_0(t) \;=\; \int_{(0,\infty)} \beta_0^{-1} \sum_{k=0}^{\infty} (-1)^k \frac{\left(\lambda t^{\alpha+1}\right)^{A_1^k}}{\prod_{i=1}^{k} A_i^k} \beta_k \, dW(\lambda), \qquad t > 0,$$

where $\beta_k = \lim_{n^* \to \infty} \frac{\Gamma\left(A_{k+1}^{n^*}+1\right)}{\prod_{i=k+1}^{n^*} A_{k+1}^i}$, $k \in \mathbb{N}_0$. Again, these results verify the existence of GOS-processes w. r. t. a class of sequences which are not necessarily eventually constant nor nonnegative like for instance sequences such that $\alpha_i = -q^i$, $i \in \mathbb{N}$, for some $0 \le q < 1$.

Lastly, in Section 5.4 we showed a way to obtain generators w. r. t. periodic sequences as solution of a system of differential equations assuming some basic structure for the functions φ_n. Since we did not study the solvability of such differential equations subject to positivity and integrability conditions, we still have no general knowledge if and w. r. t. which periodic sequences there exist GOS-processes. However, for the studied examples $\alpha_{2i-1} = 0$ and $\alpha_{2i} = 1$ resp. $\alpha_{2i} = -1$, $i \in \mathbb{N}$, we succeeded to show that corresponding GOS-processes exist by constructing appropriate generators.

Altogether, w. r. t. eventually constant parametrizing sequences we showed the existence and characterized corresponding GOS-processes. In the cases beyond eventually constant parametrizing sequences we cannot any longer characterize corresponding GOS-processes but only assure their existence in the above special cases, that is w. r. t. nonnegative parametrizing sequences, w. r. t. sequences satisfying conditions (R1) and (R2) of Section 5.3 and w. r. t. the mentioned two-periodic sequences alternating between 0 and 1 resp. between 0 and -1.

Chapter 6

Generalized order statistic processes within other classes of point processes

The present chapter's concern is the intrinsic question whether generalized order statistic processes belong to other classes of point processes or not, looking especially at two of them – on one hand we study GOS-processes as birth processes and characterize them via a recursion holding for their birth rates, on the other hand, we study the question whether GOS-processes are Cox processes or not.

6.1 Generalized order statistic processes as birth processes

Consider a (Markovian) nontrivial GOS-process N w. r. t. $\{\alpha_i\}_{i \in \mathbb{N}}$. Then N is a birth process according to Definition 8:
Firstly, for every $t > 0$ and $n \in \mathbb{N}_0$ such that $P(N_t = n) > 0$ we find

$$P(N_{t+h} > n + 1 | N_t = n) \tag{6.1}$$

$$= \frac{P(T_1 < \infty) \int_0^t \int_{s_1}^t \cdots \int_{s_{n-1}}^t \int_t^{t+h} \int_{s_{n+1}}^{t+h} f_{T_1,\dots,T_{n+2}}(s_1,\dots,s_{n+2}) \, ds_{n+2} \cdots ds_1}{P(N_t = n)}$$

$$\stackrel{Thm.32}{=} \frac{P(T_1 < \infty) \int_0^t \int_{s_1}^t \cdots \int_{s_{n-1}}^t \int_t^{t+h} \int_{s_{n+1}}^{t+h} \prod_{i=1}^{n+2} s_i^{\alpha_i} \varphi_{n+2}(s_{n+2}) \, ds_{n+2} \cdots ds_1}{P(N_t = n)}$$

$$= c(t) \int_t^{t+h} \int_{s_{n+1}}^{t+h} s_{n+1}^{\alpha_{n+1}} s_{n+2}^{\alpha_{n+2}} \varphi_{n+2}(s_{n+2}) \, ds_{n+2} ds_{n+1}$$

$$= o(h), \tag{6.2}$$

where c is a function independent of h. Thereby the last identity holds because $s_{n+1}^{\alpha_{n+1}} s_{n+2}^{\alpha_{n+2}} \varphi_{n+2}(s_{n+2})$ is bounded on e.g. $[t, t+x]^2$ for arbitrary $x > 0$.

Secondly, N satisfies $P(N_{t+h} = n + 1 | N_t = n) = \kappa_n(t)h + o(h)$ with birth rates κ_n, where

$$
\begin{aligned}
\kappa_0(t) &= \lim_{h \to 0+0} \frac{1}{h} P(N_{t+h} = 1 | N_t = 0) \\[2mm]
&\overset{(3.23)}{=} \lim_{h \to 0+0} \frac{1}{h} \cdot \Psi_1\left(t, t + h | \alpha_1\right) \frac{P(T_1 < \infty)\varphi_1(t + h)}{P(T_1 < \infty)\varphi_0(t) + P(T_1 = \infty)} \\[2mm]
&\overset{(3.20)}{=} -\frac{P(T_1 < \infty)\dot\varphi_0(t)}{P(T_1 < \infty)\varphi_0(t) + P(T_1 = \infty)}, \qquad t > 0,
\end{aligned}
\tag{6.3}
$$

and

$$
\begin{aligned}
\kappa_n(t) &= \lim_{h \to 0+0} \frac{1}{h} P(N_{t+h} = n + 1 | N_t = n) \\[2mm]
&\overset{(3.22)}{=} \lim_{h \to 0+0} \frac{1}{h} \cdot \Psi_1\left(t, t + h | \alpha_{n+1}\right) \frac{\varphi_{n+1}(t + h)}{\varphi_n(t)} \\[2mm]
&\overset{(3.20)}{=} \lim_{h \to 0+0} \frac{1}{h} \cdot \int_t^{t+h} s^{\alpha_{n+1}} ds \cdot \frac{\varphi_{n+1}(t + h)}{\varphi_n(t)} \\[2mm]
&= \frac{t^{\alpha_{n+1}}\varphi_{n+1}(t)}{\varphi_n(t)} = -\frac{\dot\varphi_n(t)}{\varphi_n(t)}, \qquad n \in \mathbb{N},\ t > 0.
\end{aligned}
\tag{6.4}
$$

Analogous to the recursions (3.11) for φ_n and (3.24) for the probabilities $P(N_t = n)$, the subsequent formula (6.5) holds for the intensities of a GOS-process, and moreover turns out to be characteristic within the class of (regular) birth processes:

Theorem 73: Let $\{\alpha_i\}_{i \in \mathbb{N}}$ be a real sequence and N a birth process such that $P(N_t = n) > 0$ for $n \in \mathbb{N}_0$, $t > 0$. Then the following statements are equivalent:

(i) The process N is a generalized order statistic process w. r. t. $\{\alpha_i\}_{i \in \mathbb{N}}$.

(ii) The birth rates κ_n, $n \in \mathbb{N}_0$, of N are positive, differentiable and satisfy $\int_0^t \kappa_0(s)ds < \infty$ for $t > 0$ and $\int_{t_0}^\infty \kappa_n(s)ds = \infty$ for $n \in \mathbb{N}_0$ and some $t_0 > 0$. Moreover

$$
\kappa_{n+1}(t) = \kappa_n(t) - \frac{\dot\kappa_n(t)}{\kappa_n(t)} + \frac{\alpha_{n+1}}{t}, \qquad n \in \mathbb{N}_0,\ t > 0.
\tag{6.5}
$$

Remark that $P(N_t = n) > 0$, $n \in \mathbb{N}_0$, $t > 0$, make sure that the intensities κ_n of N of N are well defined for each $n \in \mathbb{N}_0$. Further, due to Remark 37 and (3.17) resp. (3.18), $P(N_t = n) > 0$ for $n \in \mathbb{N}_0$, $t > 0$ holds especially for every nonexploding GOS-process.

Proof: Recall that $p_{<\infty} = P(T_1 < \infty)$ and $p_\infty = P(T_1 = \infty)$.

$(i) \Rightarrow (ii)$: Suppose that (i) holds and denote by $\{\varphi_n\}_{n\in\mathbb{N}_0}$ the corresponding family of functions. Note that due to equations (3.17), (3.18) and $P(N_t = n) > 0$ for $n \in \mathbb{N}_0$ and $t > 0$ we have $\varphi_n(t) > 0$ for $t > 0$ and $n \in \mathbb{N}_0$. Repeated application of the recursive relations between φ_n implies

$$\dot{\kappa}_0(t) \stackrel{(6.3)}{=} \frac{d}{dt}\left(-\frac{p_{<\infty}\dot{\varphi}_0(t)}{p_{<\infty}\varphi_0(t) + p_\infty}\right) \stackrel{(3.11)}{=} \frac{d}{dt}\left(\frac{p_{<\infty}t^{\alpha_1}\varphi_1(t)}{p_{<\infty}\varphi_0(t) + p_\infty}\right)$$

$$= \frac{(p_{<\infty}\alpha_1 t^{\alpha_1-1}\varphi_1(t) + p_{<\infty}t^{\alpha_1}\dot{\varphi}_1(t))(p_{<\infty}\varphi_0(t) + p_\infty) - p_{<\infty}t^{\alpha_1}\varphi_1(t)p_{<\infty}\dot{\varphi}_0(t)}{(p_{<\infty}\varphi_0(t) + p_\infty)^2}$$

$$\stackrel{(3.11)}{=} -\frac{\alpha_1}{t}\cdot\frac{p_{<\infty}\dot{\varphi}_0(t)}{p_{<\infty}\varphi_0(t) + p_\infty} - \frac{p_{<\infty}\dot{\varphi}_0(t)}{p_{<\infty}\varphi_0(t) + p_\infty}\cdot\frac{\dot{\varphi}_1(t)}{\varphi_1(t)} + \left(\frac{p_{<\infty}\dot{\varphi}_0(t)}{p_{<\infty}\varphi_0(t) + p_\infty}\right)^2$$

$$\stackrel{(6.3),(6.4)}{=} \frac{\alpha_1}{t}\kappa_0(t) - \kappa_0(t)\cdot\kappa_1(t) + \kappa_0^2(t)$$

resp.

$$\dot{\kappa}_n(t) \stackrel{(6.4)}{=} \frac{d}{dt}\left(-\frac{\dot{\varphi}_n(t)}{\varphi_n(t)}\right) \stackrel{(3.11)}{=} \frac{d}{dt}\left(\frac{t^{\alpha_{n+1}}\varphi_{n+1}(t)}{\varphi_n(t)}\right)$$

$$= \frac{\alpha_{n+1}t^{\alpha_{n+1}-1}\varphi_{n+1}(t)\varphi_n(t) + t^{\alpha_{n+1}}\dot{\varphi}_{n+1}(t)\varphi_n(t) - t^{\alpha_{n+1}}\varphi_{n+1}(t)\dot{\varphi}_n(t)}{\varphi_n^2(t)}$$

$$\stackrel{(3.11)}{=} -\frac{\alpha_{n+1}}{t}\cdot\frac{\dot{\varphi}_n(t)}{\varphi_n(t)} - \frac{\dot{\varphi}_n(t)}{\varphi_n(t)}\cdot\frac{\dot{\varphi}_{n+1}(t)}{\varphi_{n+1}(t)} + \left(\frac{\dot{\varphi}_n(t)}{\varphi_n(t)}\right)^2$$

$$\stackrel{(6.4)}{=} \frac{\alpha_{n+1}}{t}\kappa_n(t) - \kappa_n(t)\cdot\kappa_{n+1}(t) + \kappa_n^2(t), \qquad t > 0, n \in \mathbb{N}.$$

Moreover, the intensities are positive on $(0, \infty)$:

Due to (3.11) and $\varphi_n(t) > 0$ we obtain $-\dot{\varphi}_n(t) = t^{\alpha_{n+1}}\varphi_{n+1}(t) > 0$ for $n \in \mathbb{N}_0$ and $t > 0$. Equation (6.4) then implies $\kappa_n(t) > 0$, $n \in \mathbb{N}_0$, $t > 0$.

Hence, (6.5) is verified. The integrability conditions for the intensities hold as well since

$$\int_{t_0}^\infty \kappa_n(s)ds = \int_{t_0}^\infty -\frac{\dot{\varphi}_n(s)}{\varphi_n(s)}ds = -\ln\varphi_n(s)\Big|_{t_0}^\infty = \ln\varphi_n(t_0) - \lim_{s\to\infty}\ln\varphi_n(s) = \infty,$$

analogously for $\int_{t_0}^\infty \kappa_0(s)ds = \infty$, and

$$\int_0^t \kappa_0(s)ds = -\ln\varphi_0(s)\Big|_0^t = 1 - \ln\varphi_0(t) < \infty, \qquad t, t_0 > 0.$$

$(ii) \Rightarrow (i)$: Suppose that (ii) holds and put

$$\varphi_0(t) = P(T_1 > t|T_1 < \infty) = \frac{P(N_t = 0) - p_\infty}{p_{<\infty}}, \qquad t > 0.$$

Then, φ_0 is positive and differentiable on $(0, \infty)$ since $p_\infty < P(N_t = 0)$ as κ_0 is positive, and $P(N_t = 0) = e^{-\int_0^t \kappa_0(s)ds}$, compare equation (2.9). Further

$$\kappa_0(t) = -\frac{\frac{d}{dt}P(N_t = 0)}{P(N_t = 0)} = -\frac{p_{<\infty}\dot{\varphi}_0(t)}{p_{<\infty}\varphi_0(t) + p_\infty}, \qquad t > 0.$$

Hence $-\dot{\varphi}_0$ is positive and continuous since the intensities themselves are so.

The next step is to prove by induction that φ_0 is infinitely often differentiable on $(0, \infty)$ and that the recursively defined functions

$$\varphi_i(t) = -\frac{\dot{\varphi}_{i-1}(t)}{t^{\alpha_i}}, \qquad t > 0, \tag{6.6}$$

are positive and with

$$\kappa_i(t) = -\frac{\dot{\varphi}_i(t)}{\varphi_i(t)}, \qquad t > 0, \tag{6.7}$$

for $i \in \mathbb{N}$. Suppose for some $n \in \mathbb{N}_0$ that the $(n+1)$-th derivative of φ_0 exists and is continuous on $(0, \infty)$, and that further (6.6) and (6.7) hold for $i = 1, \ldots, n$. (Note that this is satisfied especially for $n = 0$.) Then, with κ_n and φ_n also $\dot{\varphi}_n$ is differentiable, i.e. φ_0 is $(n+2)$-times differentiable, and $-\dot{\varphi}_n$ is positive on $(0, \infty)$. If $n = 0$ we have

$$\frac{\dot{\kappa}_0(t)}{\kappa_0(t)} = \frac{d}{dt}\ln \kappa_0(t) = \frac{d}{dt}\ln\big(-p_{<\infty}\dot{\varphi}_0(t)\big) - \ln\big(p_{<\infty}\varphi_0(t) + p_\infty\big) = \frac{\ddot{\varphi}_0(t)}{\dot{\varphi}_0(t)} + \kappa_0(t)$$

resp. if $n > 0$

$$\frac{\dot{\kappa}_n(t)}{\kappa_n(t)} = \frac{d}{dt}\ln \kappa_n(t) = \frac{d}{dt}\ln\big(-\dot{\varphi}_n(t)\big) - \ln \varphi_n(t) = \frac{\ddot{\varphi}_n(t)}{\dot{\varphi}_n(t)} + \kappa_n(t), \quad t > 0. \tag{6.8}$$

Equation (6.5) implies

$$\kappa_{n+1}(t) = \kappa_n(t) - \frac{\dot{\kappa}_n(t)}{\kappa_n(t)} + \frac{\alpha_{n+1}}{t} \overset{(6.8)}{=} \kappa_n(t) - \frac{\ddot{\varphi}_n(t)}{\dot{\varphi}_n(t)} - \kappa_n(t) + \frac{\alpha_{n+1}}{t}$$

$$= -\frac{\ddot{\varphi}_n(t)}{\dot{\varphi}_n(t)} + \frac{\alpha_{n+1}}{t}, \qquad t > 0.$$

Put

$$\varphi_{n+1}(t) = -\frac{\dot{\varphi}_n(t)}{t^{\alpha_{n+1}}}, \qquad t > 0,$$

and note that, thus defined, $\varphi_{n+1}(t) > 0$ for $t > 0$. Further, as $\ddot{\varphi}_n$ exists, φ_{n+1} is differentiable and for its logarithmic derivative we find

$$\frac{\dot{\varphi}_{n+1}(t)}{\varphi_{n+1}(t)} = \frac{d}{dt}\ln \varphi_{n+1}(t) = \frac{d}{dt}\big(\ln\big(-\dot{\varphi}_n(t)\big) - \ln t^{\alpha_{n+1}}\big) = \frac{\ddot{\varphi}_n(t)}{\dot{\varphi}_n(t)} - \frac{\alpha_{n+1}}{t},$$

which yields

$$\kappa_{n+1}(t) = -\frac{\dot{\varphi}_{n+1}(t)}{\varphi_{n+1}(t)}, \qquad t > 0.$$

With κ_{n+1} and φ_{n+1} the derivative of φ_{n+1} is continuous and φ_0 is $(n+2)$-times continuously differentiable.

Altogether we obtain

$$\varphi_0(t) = \frac{P(N_t = 0) - P(T_1 = \infty)}{P(T_1 < \infty)} = \frac{e^{-\int_0^t \kappa_0(s)ds} - p_\infty}{p_{<\infty}}, \qquad t > 0,$$

$$\varphi_n(t) = \varphi_n(t_0)e^{-\int_{t_0}^t \kappa_n(s)ds}, \qquad n \in \mathbb{N}, \, 0 < t_0 < t,$$

where φ_0 is such that $\lim_{t\to 0} \varphi_0(t) = 1$ since $\int_0^t \kappa_0(s)ds < \infty$ implies

$$\lim_{t\to 0} \frac{e^{-\int_0^t \kappa_0(s)ds} - p_\infty}{p_{<\infty}} = \frac{e^0 - p_\infty}{p_{<\infty}} = 1.$$

Finally, $\lim_{t\to\infty} \varphi_0(t) = 0$ respectively $\lim_{t\to\infty} \varphi_n(t) = 0$ for $n \in \mathbb{N}$ hold since $\lim_{t\to\infty} P(N_t = 0) = P(T_1 = \infty)$ resp. since $\int_{t_0}^\infty \kappa_n(s)ds = \infty$ for some $t_0 > 0$.

By construction of $\{\varphi_n\}_{n\in\mathbb{N}_0}$, there exists a point process whose distribution is such that, provided that a jump occurs, its first n successive occurrence times follow the joint densities

$$f(t_1, \ldots, t_n) = \prod_{i=1}^n t_i^{\alpha_i} \varphi_n(t_n), \qquad n \in \mathbb{N}, \, 0 < t_1 \leq \cdots \leq t_n.$$

Moreover, it satisfies the generalized order statistic property according to Theorem 32. Its intensities equal $-\frac{p_{<\infty}\dot{\varphi}_0}{p_{<\infty}\varphi_0 + p_\infty}$ resp. $-\frac{\dot{\varphi}_n}{\varphi_n}$, $n \in \mathbb{N}$, compare equations (6.4) and (6.3), and thus coincide with the intensities κ_n, $n \in \mathbb{N}_0$, of N. Note that the continuity of κ_n on $(0,\infty)$, $n \in \mathbb{N}_0$, and $\int_0^t \kappa_0(s)ds < \infty$, $t > 0$, are sufficient for the intensities κ_n, $n \in \mathbb{N}_0$, to uniquely determine the transition probabilities $P(N_t = n | N_s = m)$, $0 \leq s \leq t$, $n, m \in \mathbb{N}_0$, of the corresponding process, compare Lundberg (1964), Theorem 1 on p. 40 and the following remark on p. 47. As the distribution of a birth process is completely described by these transition probabilities and e.g. the probability $P(N_0 = 0) = 1$, the distribution of a birth process with the intensities κ_n, $n \in \mathbb{N}_0$, is unique. Hence, (ii) implies that the distribution of N is the one of a GOS-process w.r.t. $\{\alpha_i\}_{i\in\mathbb{N}}$ and the proof is complete. ∎

This result extends an analogous property of mixed Poisson processes which can be found in (Grandell, 1997, Thm. 6.1).

6.2 Dissociation from Cox processes

In Chapter 4.1 Proposition 39 we have seen that w.r.t. a constant parametrizing sequence a GOS-process is a time transformed mixed Poisson process and therefore a Cox process. Hence, looking for a class of point processes possibly including GOS-processes, Cox processes are a nearby choice. However, it is the aim of this section to show that, besides few cases, in general point processes cannot simultaneously satisfy the generalized order statistic property and be Cox processes. Thereby, in the sequel we consider eventually constant parametrizing sequences, only.

Whereas the main result of this section, Theorem 78, is presented at the end and gives a complete characterization of those among the considered GOS-processes which are Cox processes simultaneously, we approach the issue studying two rather simple examples of processes with the generalized order statistic property one of which turns out to be no Cox process whereas the other one does so:

Example 74: Consider the delayed renewal process N from Example 41 with $\alpha_1 = -\frac{1}{2}$ whose occurrence times $T_1, \ldots, T_n$ follow the density

$$f_{T_1,\ldots,T_n}(t_1, \ldots, t_n) = t_1^{-\frac{1}{2}} \cdot \frac{\lambda^{n-\frac{1}{2}}}{\Gamma(\frac{1}{2})} e^{-\lambda t_n}, \qquad n \in \mathbb{N}, 0 < t_1 \leq \cdots \leq t_n,$$

for some $\lambda > 0$. Then N is a GOS-process w. r. t. a sequence $\{\alpha_i\}_{i\in\mathbb{N}}$ such that $\alpha_1 = -\frac{1}{2}$ and $\alpha_i = 0$ for $i > 1$. In terms of Theorem 32 this corresponds to $\varphi_1(t) = \sqrt{\frac{\lambda}{\pi}} e^{-\lambda t}$, $\varphi_2(t) = \frac{\lambda^{\frac{3}{2}}}{\sqrt{\pi}} e^{-\lambda t}$, $\varphi_3(t) = \frac{\lambda^{\frac{5}{2}}}{\sqrt{\pi}} e^{-\lambda t}$ and so on and according to equation (3.17) with $\gamma_n = n - \frac{1}{2}$, $n \in \mathbb{N}$, we obtain the following probabilities:

$$P(N_t = 1) = \frac{t^{\gamma_1}}{\gamma_1} \varphi_1(t) = \frac{(\lambda t)^{\frac{1}{2}}}{\frac{1}{2} \cdot \sqrt{\pi}} e^{-\lambda t},$$

$$P(N_t = 2) = \frac{t^{\gamma_2}}{\gamma_1 \gamma_2} \varphi_2(t) = \frac{(\lambda t)^{\frac{3}{2}}}{\frac{1}{2} \cdot \frac{3}{2} \cdot \sqrt{\pi}} e^{-\lambda t},$$

$$P(N_t = 3) = \frac{t^{\gamma_3}}{\gamma_1 \gamma_2 \gamma_3} \varphi_3(t) = \frac{(\lambda t)^{\frac{5}{2}}}{\frac{1}{2} \cdot \frac{3}{2} \cdot \frac{5}{2} \cdot \sqrt{\pi}} e^{-\lambda t}, \qquad t > 0.$$

If N was a Cox process, for fixed t the state N_t would be mixed Poisson distributed, hence there would exist mixing distributions V_t on $[0, \infty)$ such that

$$P(N_t = n) = \int_{[0,\infty)} \frac{x^n}{n!} e^{-x} \, dV_t(x), \qquad n \in \mathbb{N}_0, t > 0. \tag{6.9}$$

The question whether such mixing distributions exist leads to Stieltjes' moment problem, compare Section B.3:

Suppose the state N_t at time $t > 0$ would be mixed Poisson distributed with mixing distribution V_t, that is (6.9) would hold. Let $\tilde{V}_t$ be the measure (not necessarily a probability measure) which is absolutely continuous with respect to V_t such that $\frac{d\tilde{V}_t}{dV_t}(x) = e^{-x}$, V_t-almost surely. Then we had the following identity for the moments of $\tilde{V}_t$:

$$\int_{[0,\infty)} x^n \, d\tilde{V}_t(x) = n! P(N_t = n), \qquad n \in \mathbb{N}_0. \tag{6.10}$$

Put $\mu_n = n!P(N_t = n)$ for $n \in \mathbb{N}_0$. Necessary for the existence of a measure $\tilde{V}_t$ such that the above equations (6.10) hold is the positivity of $\{\mu_n\}_{n=1}^{\infty}$ in the sense of Definition B.6, compare also Theorem B.7. In the present example we find

$$\mu_1 = \frac{2}{\sqrt{\pi}}(\lambda t)^{\frac{1}{2}}e^{-\lambda t}, \qquad \mu_2 = \frac{8}{3\sqrt{\pi}}(\lambda t)^{\frac{3}{2}}e^{-\lambda t}, \qquad \mu_3 = \frac{16}{5\sqrt{\pi}}(\lambda t)^{\frac{5}{2}}e^{-\lambda t}$$

for a given $t > 0$. The nonnegative polynomials $p_a(x) = (x - a)^2 = x^2 - 2ax + a^2$, for $a \in \mathbb{R}$, have the following moments with respect to the sequence $\{\mu_n\}_{n=1}^{\infty}$:

$$\mu(p_a) = \mu_3 - 2a\mu_2 + a^2\mu_1$$
$$= \frac{16}{5\sqrt{\pi}}(\lambda t)^{\frac{5}{2}}e^{-\lambda t} - 2a\frac{8}{3\sqrt{\pi}}(\lambda t)^{\frac{3}{2}}e^{-\lambda t} + a^2\frac{2}{\sqrt{\pi}}(\lambda t)^{\frac{1}{2}}e^{-\lambda t}.$$

Those are nonnegative if and only if

$$\frac{8}{5}(\lambda t)^2 - \frac{8}{3}a\lambda t + a^2 \geq 0.$$

If we choose especially $a = \lambda t$ we obtain

$$\frac{8}{5}(\lambda t)^2 - \frac{8}{3}a\lambda t + a^2 = \frac{8}{5}(\lambda t)^2 - \frac{8}{3}(\lambda t)^2 + (\lambda t)^2 = -\frac{1}{15}(\lambda t)^2 < 0.$$

Hence, for $\lambda > 0$ and $t > 0$ the moment $\mu(p_{\lambda t})$ is negative and due to Theorem B.7 N_t is not mixed Poisson distributed. The process N cannot be a Cox process. $\square$

A special case is present if the first parameter α_1 is nonnegative and the only one to be distinct from 0. Before proving that such processes actually are Cox processes, we begin to exemplarily observe that, in comparison with Example 74, in this case the one-dimensional distributions are mixed Poisson distributions:

Example 75: Consider again the delayed renewal process N from Example 41 satisfying the generalized order statistic property this time w.r.t. $\{\alpha_i\}_{i \in \mathbb{N}}$ such that $\alpha_1 = 1$ and $\alpha_i = 0$ for $i > 1$ and whose occurrence times $T_1, \ldots, T_n$ follow the densities

$$f_{T_1,\ldots,T_n}(t_1,\ldots,t_n) = t_1 \cdot \lambda^{n+1}e^{-\lambda t_n}, \qquad n \in \mathbb{N}, 0 < t_1 \leq \cdots \leq t_n,$$

for some $\lambda > 0$. With regard to Theorem 32, this yields $\varphi_1(t) = \lambda^2 e^{-\lambda t}$ resp. $\varphi_n(t) = \lambda^{n+1}e^{-\lambda t}$, $n \in \mathbb{N}$, and we find

$$P(N_t = 0) = \varphi_0(t) \overset{(3.10)}{=} \int_t^{\infty} s\lambda^2 e^{-\lambda s}\,ds = (\lambda t + 1)e^{-\lambda t},$$

$$P(N_t = n) = \frac{t^{\gamma_n}}{\prod_{i=1}^n \gamma_i}\varphi_n(t) = \frac{(\lambda t)^{n+1}}{(n+1)!}e^{-\lambda t}, \qquad n \in \mathbb{N}, t > 0.$$

As we have already seen in the previous example, the state of N at time $t > 0$ is mixed Poisson distributed if the sequence $\{\mu_n\}_{n \in \mathbb{N}_0}$, where $\mu_n = n!P(N_t = n)$, is a sequence of moments. A sufficient condition is the positivity of $\{\mu_n\}_{n=0}^{\infty}$ and $\{\mu_n\}_{n=1}^{\infty}$, compare Definition B.6 and Theorem B.7. In our example we have for a fixed $t > 0$

$$\mu_0 = (\lambda t + 1)e^{-\lambda t} \quad \text{and} \quad \mu_n = \frac{(\lambda t)^{n+1}}{n + 1}e^{-\lambda t}, \quad n \in \mathbb{N}.$$

Consider a nonnegative polynomial $p(x) = \sum_{i=0}^{k} a_i x^i$ with real coefficients and $k \in \mathbb{N}_0$, and note that $a_0 \geq 0$ since $p(0) = a_0$. The moment $\mu^0(p)$ of p with respect to $\{\mu_n\}_{n=0}^{\infty}$ is nonnegative since

$$\mu^0(p) = \sum_{i=0}^{k} a_i \mu_i = a_0 e^{-\lambda t} + \sum_{i=0}^{k} a_i \frac{(\lambda t)^{i+1}}{i + 1}e^{-\lambda t}$$

$$\geq e^{-\lambda t} \sum_{i=0}^{k} a_i \frac{(\lambda t)^{i+1}}{i + 1} = e^{-\lambda t} \int_0^{\lambda t} \sum_{i=0}^{k} a_i s^i \, ds = e^{-\lambda t} \int_0^{\lambda t} p(s) \, ds \quad \geq 0.$$

Analogous arguments hold for the moment with respect to $\{\mu_n\}_{n=1}^{\infty}$:

$$\mu^1(p) = \sum_{i=0}^{k} a_i \mu_{i+1} = \sum_{i=0}^{k} a_i \frac{(\lambda t)^{i+2}}{i + 2}e^{-\lambda t}$$

$$= e^{-\lambda t} \int_0^{\lambda t} \sum_{i=0}^{k} a_i s^{i+1} \, ds = e^{-\lambda t} \int_0^{\lambda t} s \cdot p(s) \, ds \geq 0.$$

Hence, there exists a distribution $\tilde{V}_t$ such that

$$n!P(N_t = n) = \mu_n = \int_{[0,\infty)} x^n \, d\tilde{V}_t(x), \quad n \in \mathbb{N}_0.$$

With V_t given by $\frac{dV_t}{d\tilde{V}_t}(x) = e^x$, which is a probability distribution since

$$\int_{[0,\infty)} dV_t(x) = \int_{[0,\infty)} e^x d\tilde{V}_t(x) = \sum_{n=0}^{\infty} \int_{[0,\infty)} \frac{x^n}{n!} d\tilde{V}_t(x)$$

$$= \sum_{n=0}^{\infty} \frac{\mu_n}{n!} = e^{-\lambda t} + \sum_{n=0}^{\infty} \frac{1}{n!} \cdot \frac{(\lambda t)^{n+1}}{n + 1}e^{-\lambda t} = 1$$

(due to monotone convergence), we obtain

$$P(N_t = n) = \int_{[0,\infty)} \frac{x^n}{n!}e^{-x} \, dV_t(x), \quad n \in \mathbb{N}_0.$$

Therefore, N_t is mixed Poisson distributed with mixing distribution V_t.

Note that the same arguments go through if the distribution of N is mixed with respect to λ, that is, if N is a GOS-process w.r.t. 1, 0, 0, ... and with $\varphi_0(t) = \int_{(0,\infty)} (\lambda t + 1) e^{-\lambda t} \, dW(\lambda)$, $t \geq 0$, for some probability distribution W on $(0, \infty)$. □

Let us introduce some notations before stating a proposition which affirms the precedent example:

For $s \geq 0$, $l > 0$ we define intensity functions Λ_s^l by $\Lambda_s^l(x) = l \cdot (x - s) \mathbb{1}_{(s,\infty)}(x)$, $x \geq 0$. Let $\mathcal{L}_1$ be the set of all these intensity functions, that is put

$$\mathcal{L}_1 = \left\{ \Lambda_s^l \,\middle|\, s \geq 0, \, l > 0 \right\}. \tag{6.11}$$

Consider further the canonical mapping $\pi : [0, \infty) \times (0, \infty) \to \mathcal{L}$ with $\pi(s, l) = \Lambda_s^l$, where $\mathcal{L}$ is the set of nondecreasing, continuous functions starting in $L(0) = 0$ endowed with the σ-algebra $\mathcal{H}(\mathcal{L})$, compare Section 2.2.3. Then π is $\mathcal{B}^2$-$\mathcal{H}(\mathcal{L})$-measurable since the preimage of sets $\{L \in \mathcal{L} | L(x) \leq y\}$, $x, y \geq 0$, generating $\mathcal{H}(\mathcal{L})$ is measurable:

$$\pi^{-1}\left(\{L \in \mathcal{L} | L(x) \leq y\}\right) = \left\{(s, l) \in [0, \infty) \times (0, \infty) | l(x - s) \mathbb{1}_{(s,\infty)}(x) \leq y\right\} \in \mathcal{B}^2.$$

Let a be a positive parameter. Further, let S and L be two nonnegative real random variables such that L follows some probability distribution W on $(0, \infty)$ and, given $L = l$, either S follows a $\Gamma(l, \alpha)$-distribution if $a > 0$, i.e. S and L follow the $\ell \otimes W$-density

$$\frac{l^a \, s^{\alpha-1}}{\Gamma(a)} e^{-ls} \mathbb{1}_{(0,\infty)^2}(s, l), \qquad s, l \in \mathbb{R},$$

or S is concentrated in 0 if $a = 0$. Denote by $P^{W,a}$ the probability measure on $(\mathcal{L}, \mathcal{H}(\mathcal{L}))$ induced by π w.r.t. the distribution of (S, L).

We are now able to state:

Proposition 76: *Let N be a generalized order statistic process w.r.t. $\alpha_1 \geq 0$ and $\alpha_i = 0$ for $i \geq 2$, i.e., given $T_1 < \infty$, successive occurrence times $T_1, \ldots, T_n$ follow the joint density*

$$f_{T_1,\ldots,T_n}(t_1,\ldots,t_n) = \int_{(0,\infty)} \frac{\lambda^{\alpha_1+n} \, t_1^{\alpha_1}}{\Gamma(\alpha_1+1)} e^{-\lambda t_n} \, dW(\lambda), \quad n \in \mathbb{N}, \, 0 < t_1 \leq t_2 \leq \ldots \leq t_n,$$

(6.12)

for some distribution W on $(0, \infty)$.

Then N is distributed like a Cox process whose mixing distribution is itself a mixture of P^{W,α_1} and δ_0, precisely

$$P(T_1 < \infty) P^{W,\alpha_1} + P(T_1 = \infty) \delta_0,$$

where by δ_0 we mean the measure on $(\mathcal{L}, \mathcal{H}(\mathcal{L}))$ that is concentrated in the intensity function constant 0.

Proof: If $\alpha_1 = 0$ then M is a mixed Poisson process and in particular a Cox process. Since $P^{W,0}$ is concentrated on $\mathcal{L}_0 = \{\Lambda_0^l | l > 0\}$ and especially equals the distribution of a Poisson process mixed w.r.t. W, the statement is correct.

Let further $\alpha_1 > 0$. Then, successive occurrence times of a nonstationary Poisson process $N^{s,l}$ with intensity function Λ_s^l, $s \geq 0$, $l > 0$, follow the joint density

$$f_{T_1^{s,l},\ldots,T_n^{s,l}}(t_1,\ldots,t_n) = l^n e^{-l(t_n-s)} \mathbb{1}_{(s,\infty)}(t_1), \qquad n \in \mathbb{N}, \, 0 < t_1 \leq \cdots \leq t_n,$$

since the process M given by $M_t = N^{s,l}(t+s)$ who jumps at times $T_i^{s,l} - s$, $i \in \mathbb{N}$, is a stationary Poisson process with intensity l. This yields for the density of n successive occurrence times of a Cox process with mixing distribution P^{W,α_1}

$$f(t_1,\ldots,t_n) = \int_{\mathcal{L}} f_{T_1^\Lambda,\ldots,T_n^\Lambda}(t_1,\ldots,t_n) \, dP^{W,\alpha_1}(\Lambda)$$

$$= \int_{(0,\infty)} \int_0^\infty \lambda^n e^{-\lambda(t_n-s)} \mathbb{1}_{(s,\infty)}(t_1) \frac{\lambda^{\alpha_1}}{\Gamma(\alpha_1)} s^{\alpha_1-1} e^{-\lambda s} \, ds \, dW(\lambda)$$

$$= \int_{(0,\infty)} \frac{\lambda^{\alpha_1+n}}{\Gamma(\alpha_1)} e^{-\lambda t_n} \int_0^{t_1} s^{\alpha_1-1} \, ds \, dW(\lambda)$$

$$= \int_{(0,\infty)} \frac{\lambda^{\alpha_1+n}}{\Gamma(\alpha_1+1)} t_1^{\alpha_1} e^{-\lambda t_n} \, dW(\lambda), \qquad n \in \mathbb{N}, \, 0 < t_1 \leq \cdots \leq t_n,$$

where $T_1^\Lambda, T_2^\Lambda, \ldots$ denote successive occurrence times of a Poisson process with intensity function Λ. If N is such that at least one jump occurs almost surely, i.e. $P(T_1 < \infty) = 1$, than its distribution equals that of the considered Cox process. Otherwise, if N is such that $P(T_1 = \infty) > 0$ then there exists a GOS-process $\tilde{N}$ w.r.t. the given parametrizing sequence such that $\tilde{N}$ jumps almost surely at least once and such that its occurrence times follow the densities (6.12), compare Lemma 29 and the subsequent Remark 30. The distribution of N, provided that $T_1 < \infty$, coincides with that one of $\tilde{N}$ and we find

$$P_N = P(T_1 < \infty)P_{\tilde{N}} + P(T_1 = \infty)P^0$$

$$= P(T_1 < \infty) \int_{\mathcal{L}_1} P^\Lambda dP^{W,\alpha_1}(\Lambda) + P(T_1 = \infty)P^0 = \int_{\mathcal{L}} P^\Lambda d\tilde{P}^{W,\alpha_1}(\Lambda),$$

where P^0 denotes the distribution of a process remaining in its initial state 0 almost surely, where further P^Λ denotes the distribution of a nonstationary Poisson process with intensity function Λ and where $\tilde{P}^{W,\alpha_1} = P(T_1 < \infty)P^{W,\alpha_1} + P(T_1 = \infty)\delta_0$ is a probability measure on $(\mathcal{L}, \mathcal{H}(\mathcal{L}))$. ∎

General case

Point processes which at the same time satisfy the generalized order statistic property and are Cox processes as described in Proposition 76 are an exceptional case. This sections project is to give a complete characterization of those GOS-processes w. r. t. eventually constant sequences, which are Cox processes at the same time.

Let us begin with a preliminary lemma concerning the distribution of shifted Cox processes: For a point process N with occurrence times $T_1, T_2, \ldots$ we denote by $N^{[k]}$, $k \in \mathbb{N}$, the point process shifted by k points, that is with occurrence times $T_1^{[k]}, T_2^{[k]}, \ldots$ verifying $T_i^{[k]} = T_{i+k} - T_k$ if $T_{i+k} < \infty$ or $T_i^{[k]} = \infty$ else, $i \in \mathbb{N}$. Denote by $\mathcal{L}^\uparrow \subset \mathcal{L}$ the set of unbounded intensity functions, i. e. put

$$\mathcal{L}^\uparrow = \{\Lambda \in \mathcal{L} |\, \lim_{t \to \infty} \Lambda(t) = \infty\}.$$

For an intensity function Λ and $s > 0$ denote by $\Lambda(\cdot + s)$ the intensity function such that $\Lambda(\cdot + s)(t) = \Lambda(t + s) - \Lambda(s)$, $t \geq 0$.

Lemma 77: *Let N be a Cox process with a mixing distribution V concentrated on $\mathcal{L}^\uparrow$. Then the distribution of $N^{[k]}$ satisfies*

$$P_{N^{[k]}}(A) = \int\limits_{\Lambda \in \mathcal{L}} \int\limits_{s \in (0,\infty)} \frac{\Lambda(s)^{k-1}}{(k-1)!} e^{-\Lambda(s)} \mathrm{P}^{\Lambda(\cdot+s)}(A) \, d\Lambda(s) \, dV(\Lambda), \quad A \in \mathcal{H}(\mathcal{N}), \quad (6.13)$$

hence $N^{[k]}$ is a Cox process with mixing distribution $V^{[k]}$ such that

$$V^{[k]}(B) = \int\limits_{\Lambda \in \mathcal{L}} \int\limits_{s \in (0,\infty)} \frac{\Lambda(s)^{k-1}}{(k-1)!} e^{-\Lambda(s)} \mathbb{1}_B(\Lambda(\cdot+s)) \, d\Lambda(s) \, dV(\Lambda), \quad B \in \mathcal{H}(\mathcal{L}). \quad (6.14)$$

Proof: Since V is concentrated on $\mathcal{L}^\uparrow$ we find $P(T_k < \infty) = 1$. Denote by D_k the mapping that maps N onto $N^{[k]}$, $k \in \mathbb{N}$. Then

$$P_{N^{[k]}}(A) = P\left(N^{[k]} \in A\right) = P\left(N \in D_k^{-1}(A)\right) = P_N\left(D_k^{-1}(A)\right)$$

$$= \int\limits_{\Lambda \in \mathcal{L}} P^\Lambda\left(D_k^{-1}(A)\right) dV(\Lambda) = \int\limits_{\Lambda \in \mathcal{L}} P\left(N_\Lambda^{[k]} \in A\right) dV(\Lambda), \quad A \in \mathcal{H}(\mathcal{N}),$$

where P^Λ is the distribution of an inhomogeneous Poisson process with intensity function Λ, $N_\Lambda \sim P^\Lambda$ such a Poisson process and $N_\Lambda^{[k]}$ the process shifted by k points. We have

$$P\left(N_\Lambda^{[k]} \in A\right) = \int_{s \in (0,\infty)} P\left(N_\Lambda^{[k]} \in A \middle| T_k = s\right) dP_{T_k}(s)$$

$$\overset{(2.20)}{=} \int_{s \in (0,\infty)} P\left(N_\Lambda^{[k]} \in A \middle| T_k = s\right) \frac{\Lambda(s)^{k-1}}{(k-1)!} e^{-\Lambda(s)} \, d\Lambda(s), \quad A \in \mathcal{H}(\mathcal{N}),$$

and thus equation (6.13) holds since $N_\Lambda^{[k]}$, given $T_k = s$, is again a Poisson process with shifted intensity function $\Lambda(\cdot + s)$. Further, equation (6.13) can be interpreted as mixture

$$P_{N^{[k]}}(A) = \int\limits_{\mathcal{L}\times(0,\infty)} \mathrm{P}^{\Lambda(\cdot+s)}(A)\, d\nu(\Lambda, s), \quad A \in \mathcal{H}(\mathcal{N}), \qquad (6.15)$$

w. r. t. a probability measure ν on $\mathcal{H}(\mathcal{L}) \otimes \mathcal{B}_+$, where $\mathcal{B}_+$ denotes the Borelian σ-algebra on $(0,\infty)$, such that

$$\nu(B) = \int\limits_{\Lambda\in\mathcal{L}} \int\limits_{s\in(0,\infty)} \frac{\Lambda(s)^{k-1}}{(k-1)!} e^{-\Lambda(s)} d\Lambda(s) dV(\Lambda), \quad B \in \mathcal{H}(\mathcal{L}) \otimes \mathcal{B}_+.$$

Applying Proposition C.4 to the integral (6.15), where in this case $T : \mathcal{L}\times(0,\infty) \to \mathcal{L}$ maps (Λ, s) onto $\Lambda(\cdot + s)$, yields

$$P_{N^{[k]}}(A) = \int\limits_{\mathcal{L}} \mathrm{P}^{\Lambda}(A)\, dT(\nu)(\Lambda), \quad A \in \mathcal{H}(\mathcal{N}). \qquad (6.16)$$

Since the induced measure $T(\nu)$ satisfies

$$T(\nu)(B) = \nu(T^{-1}(B))$$

$$= \int\limits_{\Lambda\in\mathcal{L}} \int\limits_{s\in(0,\infty)} \frac{\Lambda(s)^{k-1}}{(k-1)!} e^{-\Lambda(s)} \mathbb{1}_B(\Lambda(\cdot + s)) d\Lambda(s) dV(\Lambda), \quad B \in \mathcal{H}(\mathcal{L}),$$

that is, (6.14) holds. ∎

We can now state our main result concerning the question whether generalized order statistic processes are Cox processes or not:

Theorem 78: *Let N be a nontrivial generalized order statistic process w. r. t. a sequence such that $\alpha_i = \alpha$ for $i > n^*$, some $\alpha > -1$ and some $n^* \in \mathbb{N}$. Then, the following statements hold:*

(i) The process N is a Cox process if and only if $n^* = 1$ and $\alpha_1 \geq \alpha$.

(ii) If (i) holds and hence N is such that the corresponding functions φ_n verify

$$\varphi_n(t) = \frac{\prod_{i=1}^n \gamma_i}{\Gamma(\gamma_n + 1)} \int_{(0,\infty)} \lambda^{\frac{\gamma_n}{\alpha+1}} e^{-\lambda t^{\alpha+1}}\, dW(\lambda), \qquad n \in \mathbb{N},\, t > 0, \qquad (6.17)$$

for some distribution W on $(0,\infty)$, then the mixing distribution corresponding to N as a Cox process is

$$P(T_1 < \infty)T^{\alpha+1}\left(P^{W,\frac{\alpha_1+1}{\alpha+1}-1}\right) + P(T_1 = \infty)\delta_0,$$

where $T^{\alpha+1}\left(P^{W,\frac{\alpha_1+1}{\alpha+1}-1}\right)$ is the measure on $(\mathcal{L}, \mathcal{H}(\mathcal{L}))$ induced w. r. t. $P^{W,\frac{\alpha_1+1}{\alpha+1}-1}$ by the mapping $T^{\alpha+1} : \mathcal{L} \to \mathcal{L}$ such that $T^{\alpha+1}(\Lambda)(t) = \Lambda(t^{\alpha+1})$ for $\Lambda \in \mathcal{L}$ and $t \geq 0$ and where δ_0 is as in Proposition 76.

Proof: Consider a GOS-process N w.r.t. a sequence such that $\alpha_i = \alpha > -1$ for $i > n^*$ and some $n^* \in \mathbb{N}$. Without loss of generality, let firstly N be such that $P(T_1 < \infty) = 1$ holds. Then, if N was a Cox process, its corresponding mixing distribution would be concentrated on $\mathcal{L}^\dagger$ since if not we would have

$$P(T_1 = \infty) = \lim_{t \to \infty} P(N_t = 0) = \lim_{t \to \infty} \int_{\mathcal{L}} e^{-\Lambda(t)} dV(\Lambda) = \int_{\mathcal{L}\backslash\mathcal{L}^\dagger} e^{-\Lambda_\infty} dV(\Lambda),$$

where $\Lambda_\infty = \lim_{t \to \infty} \Lambda(t) < \infty$ for $\Lambda \in \mathcal{L}\backslash\mathcal{L}^\dagger$, and $P(T_1 = \infty)$ would exceed 0.

We begin to consider the case $\alpha = 0$. First, note that the process $N^{[n^*]}$, which we obtain shifting N by n^* occurrence times, is a mixed Poisson process: Since its sojourn times $S_k^{[n^*]}$ verify $S_k^{[n^*]} = S_{n^*+k}$ and since due to Theorem 40

$$\varphi_{n^*+k}(t) = \frac{\prod_{i=1}^{n^*+k} \gamma_i}{\Gamma(\gamma_{n^*+k}+1)} \int_{(0,\infty)} \lambda^{\gamma_{n^*+k}} e^{-\lambda t} dW(\lambda)$$

$$= \frac{\prod_{i=1}^{n^*} \gamma_i}{\Gamma(\gamma_{n^*}+1)} \int_{(0,\infty)} \lambda^{\gamma_{n^*}+k} e^{-\lambda t} dW(\lambda), \qquad k \in \mathbb{N}_0,\, t > 0,$$

for some distribution W on $(0,\infty)$, we obtain

$$f_{S_1^{[n^*]},\ldots,S_k^{[n^*]}}(s_{n^*+1},\ldots,s_{n^*+k})$$

$$\overset{(3.16)}{=} \int_0^\infty \cdots \int_0^\infty \prod_{i=1}^{n^*+k} \left(\sum_{j=1}^i s_j\right)^{\alpha_i} \varphi_{n^*+k}\left(\sum_{j=1}^{n^*+k} s_j\right) ds_1 \cdots ds_{n^*}$$

$$= \int_{(0,\infty)} \underbrace{\int_0^\infty \cdots \int_0^\infty \prod_{i=1}^{n^*} \left(\sum_{j=1}^i s_j\right)^{\alpha_i} \frac{\lambda^{\gamma_{n^*}} \prod_{i=1}^{n^*} \gamma_i}{\Gamma(\gamma_{n^*}+1)} e^{-\lambda \sum_{j=1}^{n^*} s_j} ds_1 \cdots ds_{n^*}}_{=1} \quad (6.18)$$

$$\times\ \lambda^k e^{-\lambda \sum_{j=1}^k s_{n^*+j}} dW(\lambda)$$

$$= \int_{(0,\infty)} \prod_{j=1}^k \lambda e^{s_{n^*+j}} dW(\lambda), \qquad s_{n^*+1},\ldots,s_{n^*+k} > 0, \qquad (6.19)$$

where the integral in (6.18) integrates to 1 since the integrand is a density of n^* sojourn times of a GOS-process w.r.t. the given sequence $\{\alpha_i\}_{i\in\mathbb{N}}$, compare equation (3.16) and Theorem 40. Formula (6.19) yields that $N^{[n^*]}$ is a mixed Poisson process with mixing distribution W.

Seen as a Cox process, the mixing distribution $W^{[n^*]}$ of $N^{[n^*]}$ is concentrated on the set $\mathcal{L}_0$ corresponding to scaled Lebesgue measures, $\mathcal{L}_0 = \{\Lambda_0^l | l > 0\}$. Now, if the original process was a Cox process whose mixing distribution on $\mathcal{L}^\dagger$ we denote by $\tilde{W}$ we would have due to equation (6.14) of the last lemma

$$0 = W^{[n^*]}(\mathcal{L}\backslash\mathcal{L}_0) = \int_{\Lambda \in \mathcal{L}} \int_{s \in (0,\infty)} \frac{\Lambda(s)^{n^*-1}}{(n^*-1)!} e^{-\Lambda(s)} \mathbb{1}_{\mathcal{L}\backslash\mathcal{L}_0}(\Lambda(\cdot + s)) \, d\Lambda(s) \, d\tilde{W}(\Lambda).$$

This is true if and only if

$$0 = \int_{s \in (0,\infty)} \frac{\Lambda(s)^{n^*-1}}{(n^*-1)!} e^{-\Lambda(s)} \mathbb{1}_{\mathcal{L} \backslash \mathcal{L}_0}(\Lambda(\cdot + s)) \, d\Lambda(s) \qquad (6.20)$$

for $\tilde{W}$-almost every Λ. For Λ such that $\Lambda(\cdot + s) \in \mathcal{L}^\uparrow \backslash \mathcal{L}_0$ for all $s \geq 0$ we have

$$\mathbb{1}_{\mathcal{L} \backslash \mathcal{L}_0}(\Lambda(\cdot + s)) = 1, \qquad s \geq 0,$$

and (6.20) cannot hold for such a Λ. Hence, $\tilde{W}$ must be concentrated on

$$\mathcal{L}_0^* = \{\Lambda \in \mathcal{L}^\uparrow | \exists s \geq 0 : \Lambda(\cdot + s) \in \mathcal{L}_0\} = \{\Lambda \in \mathcal{L}^\uparrow | \exists s \geq 0, l > 0 : \Lambda(\cdot + s) = \Lambda_0^l(\cdot)\}.$$

Since for $\Lambda \in \mathcal{L}_0^*$ there exists $s^* \geq 0$ such that

$$\mathbb{1}_{\mathcal{L} \backslash \mathcal{L}_0}(\Lambda(\cdot + s)) = \begin{cases} 0, & s \geq s^*, \\ 1, & 0 \leq s < s^*, \end{cases}$$

equation (6.20) holds if and only if $\Lambda(s^*) = 0$ which again holds if and only if $\Lambda \in \mathcal{L}_1 \subset \mathcal{L}_0^*$, compare (6.11).

If N was a Cox process its occurrence times $T_1, \ldots, T_n$ would follow the density

$$f_{T_1,\ldots,T_n}(t_1,\ldots,t_n) = \int_{\mathcal{L}_1} f_{T_1,\ldots,T_n}^\Lambda(t_1,\ldots,t_n) \, d\tilde{W}(\Lambda)$$

$$= \int_{s \in [0,t_1)} \int_{l \in (0,\infty)} f_{T_1,\ldots,T_n}^{\Lambda_s^l}(t_1,\ldots,t_n) \, d(\pi^{-1}\tilde{W})(s,l)$$

$$= \int_{s \in [0,t_1)} \int_{l \in (0,\infty)} l^n e^{-l(t_n-s)} \, d(\pi^{-1}\tilde{W})(s,l), \qquad n \in \mathbb{N}, \, 0 < t_1 \leq t_2 \cdots \leq t_n,$$

where $f_{T_1,\ldots,T_n}^\Lambda$ resp. $f_{T_1,\ldots,T_n}^{\Lambda_s^l}$ denote the joint densities of n successive occurrence times of nonstationary Poisson processes with intensity function Λ resp. Λ_s^l and where $\pi^{-1}\tilde{W}$ is the image measure of $\tilde{W}$ with respect to the mapping $\pi^{-1} : \mathcal{L}_1 \to [0,\infty) \times (0,\infty)$ such that $\pi^{-1}(\Lambda_s^l) = (s,l)$ with $s \geq 0$ and $l > 0$. On the other hand, N as a GOS-process verifies

$$f_{T_1,\ldots,T_n}(t_1,\ldots,t_n) = \frac{\prod_{i=1}^n t_i^{\alpha_i \gamma_i}}{\Gamma(\gamma_n+1)} \int_{(0,\infty)} \lambda^{\gamma_n} e^{-\lambda t_n} \, dW(\lambda), \qquad n \geq n^*,$$

for some distribution W, compare Theorem 40, which would imply

$$\frac{\prod_{i=1}^n t_i^{\alpha_i \gamma_i}}{\Gamma(\gamma_n+1)} \int_{(0,\infty)} \lambda^{\gamma_n} e^{-\lambda t_n} \, dW(\lambda) = \int_{s \in [0,t_1)} \int_{l \in (0,\infty)} l^n e^{-l(t_n-s)} \, d(\pi^{-1}\tilde{W})(s,l), \quad n \geq n^*.$$

$$(6.21)$$

Since for $\alpha_1 < 0$ the left side of the equation is strictly decreasing whereas the right side is increasing in t_1, equation (6.21) cannot hold if $\alpha_1 < 0$ and a corresponding

GOS-process can thus not be a Cox process. Moreover, if there exists $i > 1$ with $\alpha_i \neq 0$ then, for $n \geq i + 1$, the left side of the equation depends on t_i whereas the right side does not, which again contradicts the assumption that a corresponding GOS-process is a Cox process.

Let us proceed with the more general case where α is an arbitrary real number exceeding -1 and where N is a GOS-process w. r. t. a sequence such that $\alpha_i = \alpha$ for $i > n^*$ and some $n^* \in \mathbb{N}$. The time transformed process M such that

$$M_t = N(t^c), \qquad t \geq 0,$$

where $c = (\alpha+1)^{-1}$ satisfies the generalized order statistic property w. r. t. a sequence $\{\tilde{\alpha}_i\}_{i \in \mathbb{N}}$ such that $\tilde{\alpha}_i = c \cdot (\alpha_i + 1) - 1$, compare Section 3.5.1. This yields $\tilde{\alpha}_i = 0$ for $i > n^*$. Due to the first part of the proof M is a Cox process if and only if $\tilde{\alpha}_1 \geq 0$, that is $\alpha_1 \geq \alpha$, and $\tilde{\alpha}_i = 0$, that is $\alpha_i = \alpha$, for $i > 1$. According to Proposition 76, the mixing distribution of M equals $P(T_1 < \infty) P^{W, \tilde{\alpha}_1} + P(T_1 = \infty) \delta_0$.

Since the class of Cox processes is closed with respect to continuous time transformations, N is a Cox process if and only if M is one. To be more precise, if $T^{\alpha+1} : \mathcal{L} \to \mathcal{L}$ denotes the mapping such that $T^{\alpha+1}\Lambda(t) = \Lambda(t^{\alpha+1})$ for $t \geq 0$ then due to Proposition 17 the mixing distribution of N is the measure induced by $T^{\alpha+1}$:

$$T^{\alpha+1}\left(P(T_1 < \infty) P^{W, \tilde{\alpha}_1} + P(T_1 = \infty) \delta_0 \right)$$
$$= P(T_1 < \infty) T^{\alpha+1}\left(P^{W, \frac{\alpha_1+1}{\alpha+1} - 1} \right) + P(T_1 = \infty) \delta_0,$$

where the last equality holds since to take the image of a measure is a linear operation and since $T^{\alpha+1}(\delta_0) = \delta_0$. ∎

In the light of this last result and its proof it is rather not to hope that any generalized order statistic process beyond the cases described, e.g., with respect to parametrizing sequences composed of an infinite number of different elements, is again a Cox process.

Chapter 7

Conclusion and outlook

In this thesis, we introduced the class of generalized order statistic processes and achieved several results. Central is the characterization given in Theorem 32, saying that a GOS-process w. r. t. a sequence $\{\alpha_i\}_{i\in\mathbb{N}}$ can be associated to a family $\{\varphi_n\}_{n\in\mathbb{N}}$ of functions such that

$$f_{T_1,\ldots,T_n}(t_1,\ldots,t_n) \;=\; \prod_{i=1}^{n} t_i^{\alpha_i}\varphi_n(t_n), \qquad n\in\mathbb{N},\, 0 < t_1 \leq \cdots \leq t_n.$$

We deduced several formulas which describe the distribution of the process mainly in terms of the parametrizing sequence, the family $\{\varphi_n\}_{n\in\mathbb{N}}$ and $\varphi_0(t) = \int_t^\infty s^{\alpha_1}\varphi_1(s)ds$, $t > 0$. We developed a criterion depending on the parametrizing sequence which decides explicitly whether a GOS-process will explode with a positive probability or not.

Later, we intensively studied GOS-processes w. r. t. parametrizing sequences eventually constant, that is, such that α_i is constant $\alpha > -1$ for $i > n^*$ and some $n^* \in \mathbb{N}$. In this case we were able to specify the above densities and thus in fact the functions $\{\varphi_n\}_{n\in\mathbb{N}}$ up to a mixture, i. e.

$$f_{T_1,\ldots,T_n}(t_1,\ldots,t_n) \;=\; \frac{\prod_{i=1}^{n}\gamma_i t_i^{\alpha_i}}{\Gamma\left(\frac{\gamma_n}{\alpha+1}+1\right)}\int_{(0,\infty)} \lambda^{\frac{\gamma_n}{\alpha+1}} e^{-\lambda t_n^{\alpha+1}}\, dW(\lambda)$$

for $0 < t_1 \leq \cdots \leq t_n$ and $n \geq n^*$, compare Theorem 40. Several results in the case of eventually constant parametrizing sequences followed such as a decomposition of GOS-processes w. r. t. particular sequences $\{\alpha_i\}_{i\in\mathbb{N}}$ consisting of natural numbers in terms of mixed Poisson processes after deleting some points. As a corollary we deduced an asymptotic result studying the limiting behavior of these GOS-processes. Returning to arbitrary parametrizing sequences, the basic concept of generators φ_0 has been introduced in order to specify distributions of GOS-processes w. r. t. various other parametrizing sequences. Thereby, a generator φ_0 w. r. t. $\{\alpha_i\}_{i\in\mathbb{N}}$ is a function with $\varphi_0(0) = 1$ and such that the family $\{\varphi_n\}_{n\in\mathbb{N}}$ which is recursively defined by

$$-\varphi_{n+1}(t)\, t^{\alpha_{n+1}} \;=\; \dot{\varphi}_n(t), \qquad n\in\mathbb{N}_0,\, t > 0,$$

verifies $\varphi_n(t) \geq 0$ for $n \in \mathbb{N}_0$, $t > 0$, and $\lim_{t \to \infty} \varphi_n(t) = 0$ for $n \in \mathbb{N}_0$. These conditions assure, that to put

$$f_n(t_1, \ldots, t_n) = \prod_{i=1}^{n} t_i^{\alpha_i} \varphi_n(t), \qquad n \in \mathbb{N}, 0 < t_1 \leq \cdots \leq t_n,$$

forms a projective family of probability densities. Thus a GOS-process can be constructed based on a single function φ_0, only. In general it is hard to find appropriate generators. Nevertheless, we succeeded to develop generators w.r.t. nonnegative parametrizing sequences, w.r.t. in principle increasing convergent sequences and w.r.t. periodic sequences. As generators w.r.t. a given sequence are nonunique we studied the relation between different generators w.r.t. one and the same parametrizing sequence. In a canonical setting excluding explosion we showed that two generators φ_0^P and φ_0^Q are associated to one and the same sequence if and only if the process

$$\left\{ \frac{\varphi_{N_t}^Q(t)}{\varphi_{N_t}^P(t)} \right\}_{t \geq 0}$$

is a martingale.

A last issue was dedicated to the classification of GOS-processes within other classes of point processes. Especially as birth processes, GOS-processes w.r.t. $\{\alpha_i\}_{i \in \mathbb{N}}$ are characterized by the following recursion holding for their birth rates:

$$\kappa_{n+1}(t) = \kappa_n(t) - \frac{\dot{\kappa}_n(t)}{\kappa_n(t)} + \frac{\alpha_{n+1}}{t}, \qquad n \in \mathbb{N}_0, t > 0. \tag{7.1}$$

We closed the thesis showing, subject to eventually constant parametrizing sequences, that only few GOS-processes lie within the class of Cox processes.

Despite the large variety of answered questions, there are just as many that remain unanswered. Central open issues are to clarify the question of existence of GOS-processes w.r.t. a given parametrizing sequence and to completely characterize these processes. This is equivalent to the characterization of all possible generators. To approach these problems, the author suggests the following ways:

Firstly, referring to the analytic concept of generators, look for the general solution of the infinite dimensional system of differential equations given by

$$\dot{\varphi}_n(t) = -\varphi_{n+1}(t) t^{\alpha_{n+1}}, \qquad n \in \mathbb{N}_0, t > 0, \tag{7.2}$$

respecting the positivity of φ_n, $\lim_{t \to \infty} \varphi_n(t) = 0$, $n \in \mathbb{N}_0$, and $\lim_{t \to 0} \varphi_0(t) = 1$. At least in connection with periodic parametrizing sequences and assuming, like in Section 5.4, that

$$\varphi_n(t|\alpha_1, \alpha_2, \ldots) = C_n \varphi_0(t|\alpha_{n+1}, \alpha_{n+2}, \ldots), \qquad n \in \mathbb{N}_0, t > 0,$$

this seems to be possible as the infinite-dimensional system (7.2) can be reduced to a finite-dimensional system of the form

$$\dot{\vec{\phi}}(t) \;=\; -C(t)\vec{\phi}(t), \qquad t > 0, \tag{7.3}$$

where $\vec{\phi}$ and C are given in (5.34). According to (Bronstein and Semendjajew, 1996, 1.12.6.1), we then obtain the general representation

$$\vec{\phi}(t) \;=\; \mathcal{T}\left(e^{-\int_0^t C(s)ds}\right), \qquad t > 0,$$

for the solution of (7.3), where $\mathcal{T}$ is the Dyson time ordering operator and where e^A for some square matrix A is the matrix exponential. However, it remains open if this solution is such that positivity and integrability can be ensured.

An alternative probabilistic approach is connected with the question if in analogy to mixed Poisson processes there exist GOS-processes N_λ with distributions P_{N_λ}, $\lambda \geq 0$, such that the distribution P_N of a GOS-process N satisfies

$$P_N(A) \;=\; \int_{[0,\infty)} P_{N_\lambda}(A)dV(\lambda), \qquad A \in \mathcal{H}(\mathcal{N}),$$

for some probability distribution V on $[0,\infty)$. The existence of such a representation is plausible as we already discovered a mixed structure in many results such as in Theorems 40 and 46. Presumably, a candidate for such an underlying standard process N_1 (for arbitrary $\lambda > 0$ a process distributed like N_λ should be given by the time transformation $t \mapsto \lambda t$) can be obtained as limit of sample processes based on generalized order statistics. This is true for Poisson processes at least:

Let N^n be the sample process corresponding to n ordinary order statistics $X_{1:n}, \ldots, X_{n:n}$ based on $\mathrm{U}[0,n]$, that is

$$N_t^n = \sup\{j \in \{1,\ldots,n\} | X_{j:n} \leq t\}, \qquad t \geq 0,\ n \in \mathbb{N}.$$

Then $\lim_{n\to\infty} N^n$ is a standard Poisson process (with intensity 1) since for $n \in \mathbb{N}$ the $k \leq n$ occurrence times $T_1^n, \ldots, T_k^n$ of N^n follow the density

$$f_{T_1^n,\ldots,T_k^n}(t_1,\ldots,t_k)$$

$$= \int_{t_k}^n \int_{s_{k+1}}^n \cdots \int_{s_{n-1}}^n \frac{n!}{n^n} ds_n \cdots ds_{k+1} \mathbb{1}_{K_k}(t_1,\ldots,t_k)$$

$$= \frac{n!}{n^n(n-k)!} (n-t_k)^{n-k} \mathbb{1}_{K_k}(t_1,\ldots,t_k)$$

which tends to the density

$$e^{-t_k} \mathbb{1}_{K_k}(t_1,\ldots,t_k), \qquad t_1,\ldots,t_n \in \mathbb{R},$$

of k successive occurrence times of a standard Poisson process for $n \to \infty$, see (2.13). Note, that the above limit construction of Poisson processes implies a similar and even characteristic property of mixed Poisson processes referring to mixed sample processes, compare Theorem 6.6 in Grandell (1997).

Admitting generalized order statistics in place of ordinary order statistics, that is, given a sequence $\{\alpha_i\}_{i\in\mathbb{N}}$, choosing for N^n the sample process corresponding to the random variables $X_{1:n}, \ldots, X_{n:n}$ with density

$$f(t_1, \ldots, t_n) \;=\; n^{-\gamma_n} \prod_{i=1}^{n} \gamma_i t_i^{\alpha_i}, \qquad 0 < t_1 \leq \ldots \leq t_n \leq n,$$

probably leads to GOS-processes as $n \to \infty$. However, this limit has to fail in case of parametrizing sequences with respect to which corresponding GOS-processes do not exist, compare Section 4.3, and a detailed analysis is necessary.

Another related question which concerns the structure of the class of GOS-processes is the following: Consider sequences within the space of parametrizing sequences, i. e. $\{\{\alpha_i^n\}_{i\in\mathbb{N}}\}_{n\in\mathbb{N}}$ and corresponding GOS-processes N^{α^n}, $n \in \mathbb{N}$, and ask for the limit

$$\lim_{n\to\infty} N^{\alpha^n}.$$

This is in principal already incorporated in Section 5.3 for sequences $\{\alpha_i^{n^*}\}_{i\in\mathbb{N}}$, $n^* \in \mathbb{N}$, such that $\alpha_i^{n^*} = \alpha_i$ for $i \leq n^*$ and $\alpha_i^{n^*} = \alpha_{n^*+1}$ for $i > n^*$ and where $\{\alpha_i\}_{i\in\mathbb{N}}$ is a sequence satisfying conditions (R1) and (R2) of this section. However, we did not yet study the limit of GOS-processes w. r. t. a sequence of parametrizing sequences in this general context.

Another remarkable structural aspect is connected with equation (3.24). For a GOS-process N w. r. t. $\{\alpha_i\}_{i\in\mathbb{N}}$ we find

$$\frac{d}{dt}P(N_t = n) \;=\; \frac{\gamma_n}{t}P(N_t = n) - \frac{\gamma_{n+1}}{t}P(N_t = n+1), \qquad n \in \mathbb{N}_0,\, t > 0. \quad (7.4)$$

Although N is a birth process, this recursion would also hold for a death process N^* with death rates $\rho_n(t) = \frac{\gamma_n}{t}$, $n \in \mathbb{N}_0$, $t > 0$. Hence, (7.4) holds if the transition probabilities of N^* verify for $n, m \in \mathbb{N}_0$

$$P(N^*_{t+h} = m | N^*_t = n) \;=\; \begin{cases} 1 - \rho_n(t)h + o(h), & m = n, \\ \rho_n(t)h + o(h), & m = n - 1, \\ o(h), & else, \end{cases} \qquad \text{as } h \downarrow 0.$$

In addition, the rates ρ_n, $n \in \mathbb{N}_0$, satisfy (6.4), that is

$$\rho_{n+1}(t) \;=\; \rho_n(t) - \frac{\dot{\rho}_n(t)}{\rho_n(t)} + \frac{\alpha_{n+1}}{t}, \qquad n \in \mathbb{N}_0,\, t > 0.$$

It is of great interest to understand how such a dual process N^* can be appropriately defined (what about N_0^*?) and in which way it is related to a GOS-process.

Leaving the characterization and construction of GOS-processes aside, there is a large variety of questions concerning GOS-processes w.r.t. particular parameter constellations. One might in particular think of GOS-processes w.r.t. eventually constant sequences which are completely described by Theorem 40. Inspired by the well-developed theory for mixed Poisson processes which also correspond to a particular parametrization for instance the following issues are of interest:

- Application: Find stochastic models or algorithms which lead to appropriate GOS-processes. Where can GOS-processes be successfully applied? Where do they naturally occur?

- Asymptotic properties: Deduce an analog to the asymptotic property $\lim_{n\to\infty} \frac{n}{T_n} = \Lambda$ of mixed Poisson processes, as partly done in Corollary 52.

- Mixing distributions: Consider particular (parametric) families of mixing distributions and study corresponding GOS-processes.

- Characterizations within other classes of point processes: For mixed Poisson processes there exists a variety of characterizations, in particular within general point processes. Which of them can be generalized to GOS-processes?

- Statistics: Develop appropriate methods in order to estimate e.g. the mixing distribution corresponding to a GOS-process.

A last considerable topic we want to mention aims at the results of Feigin (1979) and Puri (1982). It is the near by question which related processes occur if the conditional densities defining the generalized order statistic property do not origin in independent uniform but arbitrary other distributions. In other words, one can admit generalized order statistics based on arbitrary probability distributions in the definition of the generalized order statistic property. This will surely be connected to time transformations as it is the case for the (ordinary) order statistic property compared to the uniform order statistic property.

Appendix A

Special functions

A.1 Gamma and Beta function

Definition A.1: *The* **Gamma function** $\Gamma(z)$ *can be defined by*

$$\frac{1}{\Gamma(z)} = ze^{\gamma z} \prod_{n=1}^{\infty} \left[\left(1 + \frac{z}{n}\right) \exp\left(-\frac{z}{n}\right) \right], \qquad z \in \mathbb{C}\backslash\{0, -1, -2, \ldots\}, \qquad (A.1)$$

where γ is Euler's constant, $\gamma = \lim_{n\to\infty} \sum_{k=1}^{n} \frac{1}{k} - \ln n = 0.577...$ If z has a positive real part, $\Re(z) > 0$, this coincides with

$$\Gamma(z) = \int_0^\infty s^{z-1} e^{-s} \, ds.$$

For $z \in \mathbb{C}$ and $x \in \mathbb{R}$ such that $x > 0$ the function

$$\Gamma(z, x) = \int_x^\infty s^{z-1} e^{-s} \, ds$$

is called **incomplete Gamma function.**

The Gamma function has the poles $-\mathbb{N}_0$. It generalizes the factorial function, that is $\Gamma(z+1) = z\Gamma(z)$ holds for $z \in \mathbb{C}\backslash\{0, -1, -2, \ldots\}$. In particular, we find $\Gamma\left(\frac{1}{2}\right) = \sqrt{\pi}$. As a function on $(0, \infty)$ the Gamma function is convex and attains its only minimum at $x_0 \approx 1.46163$ where $\Gamma(x_0) \approx 0.88560$. Well known is Stirling's formula:

$$\Gamma(x + 1) = \sqrt{2\pi} x^{x+\frac{1}{2}} \exp\left\{-x + \frac{\vartheta}{12x}\right\}, \qquad x \in \mathbb{R}, \, x > 0, \qquad (A.2)$$

where ϑ is some real number such that $0 < \vartheta < 1$. The behavior of the logarithm of a Gamma function for large $|z|$ is specified by

Theorem A.2: (cp. Rainville (1960)) *As $|z| \to \infty$ in the region where $|\arg z| \leq \pi - \delta$, $\delta > 0$, we have*

$$\ln \Gamma(z) = \left(z - \frac{1}{2}\right) \ln z - z + O(1).$$

Definition A.3: *For* $z_1, z_2 \in \mathbb{C}$ *the* **Beta function** *is defined by*

$$\text{Beta}(z_1, z_2) = \int_0^1 s^{z_1-1}(1-s)^{z_2-1}\, ds.$$

We find

$$\text{Beta}(z_1, z_2) = \frac{\Gamma(z_1)\Gamma(z_2)}{\Gamma(z_1 + z_2)}, \qquad z_1, z_2 \in \mathbb{C}.$$

For a detailed account on Gamma and Beta functions we refer to the book of Rainville (1960).

A.2   Generalized hypergeometric function

Definition A.4: *For* $p, q \in \mathbb{N}_0$, $a_1, \ldots, a_p \in \mathbb{C}$ *and* $b_1, \ldots, b_q \in \mathbb{C} \backslash \{0, -1, -2, \ldots\}$ *the* **generalized hypergeometric function** *is defined by*

$$\,_p\mathrm{F}_q\left(\begin{matrix} a_1, \ldots, a_p \\ b_1, \ldots, b_q \end{matrix}\,\middle|\, z\right) = \sum_{n=0}^{\infty} \frac{\prod_{j=1}^p (a_j)_n}{\prod_{k=1}^q (b_k)_n} \cdot \frac{z^n}{n!}, \qquad z \in \mathbb{C}, \tag{A.3}$$

where $(x)_n = x \cdot (x+1) \cdots (x+n-1)$ *for* $n \in \mathbb{N}$ *and* $(x)_0 = 1$.

The domain of convergence depends on the parameters p and q. If e.g. $p \leq q$ the series on the right hand side of (A.3) converges for $z \in \mathbb{C}$ and thus it converges uniformly in compact subsets of $\mathbb{C}$. The following representation of generalized hypergeometric functions $\,_p\mathrm{F}_q$ in terms of a Mellin-Barnes integral can be found in (Rainville, 1960, Thm. 36):

Theorem A.5: *If* $\Re(z) < 0$ *and if no* a_j *or* b_k, $j, k \in \{1, \ldots, q\}$, *is zero or a negative integer, then*

$$\frac{1}{2\pi \mathrm{i}} \int_B \frac{(-z)^s \Gamma(-s) \prod_{j=1}^q \Gamma(a_j + s)\, ds}{\prod_{k=1}^q \Gamma(b_k + s)} = \frac{\prod_{j=1}^q \Gamma(a_j)}{\prod_{k=1}^q \Gamma(b_k)} \,_q\mathrm{F}_q\left(\begin{matrix} a_1, \ldots, a_q \\ b_1, \ldots, b_q \end{matrix}\,\middle|\, z\right),$$

where B *is a line from* $-\mathrm{i}\infty$ *to* $+\mathrm{i}\infty$ *curving if necessary to put the poles of* $\Gamma(-s)$ *to the right of the path and those of* $\Gamma(a_j + s)$ *and* $\Gamma(b_k + s)$ *to the left.*

The integral above is similar to an inverse Mellin transformation, compare Section B.2, the integration path B is called a Barnes path of integration.

A.3   Meijer's G-function

Definition A.6: *The following function defined via a contour integral of Mellin-Barnes type in the complex plane is called* **Meijer's G-function***:*

$$\mathrm{G}_{p,q}^{m,n}\left[z\,\middle|\,\begin{matrix} a_1, \ldots, a_p \\ b_1, \ldots, b_q \end{matrix}\right] = \frac{1}{2\pi \mathrm{i}} \int_L \frac{\prod_{j=1}^m \Gamma(b_j + s) \prod_{j=1}^n \Gamma(1 - a_j - s)}{\prod_{j=m+1}^q \Gamma(1 - b_j - s) \prod_{j=n+1}^p \Gamma(a_j + s)} z^{-s}\, ds$$

where $m, n, p, q \in \mathbb{N}_0$ with $0 \leq n \leq p$, $0 \leq m \leq q$, where the parameters $a_1, \ldots, a_p$, $b_1, \ldots, b_q \in \mathbb{C}$ are such that no poles of $\Gamma(b_j + s)$, $j = 1, \ldots, m$, coincide with any pole of $\Gamma(1 - a_j - s)$, $j = 1, \ldots, n$, and where L is a suitable contour.

Within its region of convergence which depends on the concrete choice of parameters Meijer's G-function is analytic. References where the G-function is studied in this general context are Mathai (1993) and Erdélyi et al. (1953). In the present work, Meijer's G-functions always appear in connection with a special choice of parameters why we introduce the abbreviating notation

$$G_n[z \,|\, a_1, \ldots, a_n] = G_{n,n}^{n,0}\left[z \,\Big|\, \begin{matrix} a_1, \ldots, a_n \\ a_1 - 1, \ldots, a_n - 1 \end{matrix}\right] = \frac{1}{2\pi i} \int_L \frac{z^{-s}}{\prod_{j=1}^{n} a_j - 1 + s}\, ds.$$

$$(A.4)$$

In this case, the contour L can be chosen to be a loop encircling the poles of the integrand in the positive direction and the integral converges for $z \neq 0$. For distinct values of a_i function (A.4) can be evaluated with the help of the following properties:

Lemma A.7: *For $z \neq 0$ and $n \in \mathbb{N}$ we find*

(i) $G_0[z] = 0$, $G_1[z|a_1] = z^{a_1 - 1}$,

(ii) $G_n[z|a_1, \ldots, a_n](a_n - a_1) = G_{n-1}[z|a_1, \ldots, a_{n-1}] - G_{n-1}[z|a_2, \ldots, a_n]$ and

(iii) $G_n[z|1, \ldots, n] = \frac{1}{(n-1)!}(1 - z)^{n-1}$.

Proof: These, respectively similar properties can be found (partly without proof) in Mathai (1993); Cramer (2002); Cramer and Kamps (2003). The essential steps are: for (i) (for G_0) as a function of s, z^s is holomorphic on $\mathbb{C}$, (for G_1) apply Cauchy's formula, (ii) results from the identity $\frac{a_n - a_1}{(a_1 - 1 - s)(a_n - 1 - s)} = \frac{1}{a_1 - 1 - s} - \frac{1}{a_n - 1 - s}$ and (iii) can be found by induction on n using (ii). ∎

In the context of dual generalized order statistics Burkschat et al. (2003) state the following property:

Lemma A.8: *For $n \in \mathbb{N}$ and $m_1, \ldots, m_n \in \mathbb{R}$ such that $\gamma_j = \sum_{i=1}^{j}(m_i + 1) > 0$, $j = 1, \ldots, n$, the following equation holds:*

$$\int_x^1 \int_{s_1}^1 \int_{s_2}^1 \cdots \int_{s_{n-1}}^1 \prod_{j=1}^{n} \gamma_j s_j^{m_j}\, ds_n ds_{n-1} \cdots ds_1 = 1 - \left(\prod_{j=1}^{n} \gamma_j\right) \int_0^x G_n[y \,|\, \gamma_1, \ldots, \gamma_n]\, dy.$$

A.4   Airy functions

Definition A.9: *The solutions Ai and Bi of the differential equation*

$$\frac{d^2}{dz^2} y - zy = 0, \qquad z \in \mathbb{C},$$

*such that $\mathrm{Ai}(0) = \frac{\mathrm{Bi}(0)}{\sqrt{3}} = \left(3^{\frac{2}{3}} \Gamma\left(\frac{2}{3}\right)\right)^{-1}$ and $-\mathrm{Ai}'(0) = \frac{\mathrm{Bi}'(0)}{\sqrt{3}} = \left(3^{\frac{1}{3}} \Gamma\left(\frac{1}{3}\right)\right)^{-1}$ are called **Airy functions**.*

Among various integral and series representations for Airy functions is the following which represents Ai in terms of generalized hypergeometric functions:

$$
\begin{aligned}
\mathrm{Ai}(z) &= c_1 \cdot {}_0F_1\left(\frac{2}{3}\bigg|\frac{z^3}{9}\right) - c_2 z \cdot {}_0F_1\left(\frac{4}{3}\bigg|\frac{z^3}{9}\right) \\
&= c_1 \sum_{k=0}^{\infty} 3^k \left(\frac{1}{3}\right)_k \frac{z^{3k}}{(3k)!} - c_2 \sum_{k=0}^{\infty} 3^k \left(\frac{2}{3}\right)_k \frac{z^{3k+1}}{(3k+1)!}, \qquad z \in \mathbb{C}, \quad (A.5)
\end{aligned}
$$

where $c_1 = \mathrm{Ai}(0)$ and $c_2 = \dot{\mathrm{Ai}}(0)$.

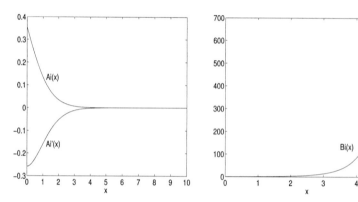

Figure A.1: Plots of $\mathrm{Ai}(x)$ and its derivative $\dot{\mathrm{Ai}}(x)$ for $x \geq 0$

Figure A.2: Plot of $\mathrm{Bi}(x)$ for $x \geq 0$

Further, Ai and Bi are real valued on $\mathbb{R}$. As the above plots indicate, if $x \geq 0$ we find $\mathrm{Ai}(x), \mathrm{Bi}(x) \geq 0$ and $-\dot{\mathrm{Ai}}(x) \geq 0$ as well as $\lim_{x\to\infty} \mathrm{Ai}(x) = \lim_{x\to\infty} \dot{\mathrm{Ai}}(x) = 0$ and $\lim_{x\to\infty} \mathrm{Bi}(x) = \infty$. For a detailed account on Airy functions we refer to Vallée and Soares (2004) and Abramowitz and Stegun (1964).

A.5   Bessel functions

Definition A.10: *For $\nu \in \mathbb{C}$ with $\Re(\nu) \geq 0$, the solutions* Bessel$_\nu$ *and* BesselK$_\nu$ *of the differential equation*

$$
z^2 \frac{d^2}{dz^2}y + z\frac{d}{dz}y - (z^2 + \nu^2)y = 0, \qquad z \in \mathbb{C}\backslash(-\infty, 0], \qquad (A.6)
$$

such that Bessel$_\nu$ *and* BesselK$_\nu$ *are independent,* Bessel$_\nu$ *is bounded as $z \to 0$ and* BesselK$_\nu$ *tends to zero as $|z| \to \infty$ in the sector $|\arg z| < \frac{\pi}{2}$ are called* **modified Bessel functions** *of the first resp. second kind.*

For $\nu > 0$ the functions $\mathrm{BesselI}_\nu$ and $\mathrm{BesselK}_\nu$ are real and positive on $(0, \infty)$. Besides $\lim_{x \to \infty} \mathrm{BesselK}_\nu(x) = 0$, we find

$$\lim_{x \to 0} \frac{\mathrm{BesselK}_\nu(x) \left(\tfrac{1}{2}x\right)^\nu}{\tfrac{1}{2}\Gamma(\nu)} = 1 \quad \text{and} \quad \lim_{x \to \infty} \mathrm{BesselI}_\nu(x) = \infty.$$

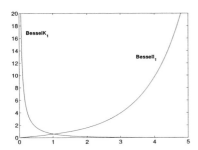

Figure A.3: Plots of modified Bessel functions of the first and second kind

The above and further properties of Bessel functions can be found in Abramowitz and Stegun (1964), note especially their formula 9.6.27 connecting the first derivative of $\mathrm{BesselK}_0$ with $\mathrm{BesselK}_1$:

$$\mathrm{BesselK}_0'(z) = -\mathrm{BesselK}_1(z), \qquad z \in \mathbb{C}\backslash(-\infty, 0]. \tag{A.7}$$

Appendix B

Integral transformations

We introduce two integral transformations, each time in the context in which it will be needed in the sequel.

B.1 Laplace transform

The following definition and subsequent theorems can essentially be found in (Feller, 1971, p. 432 f.):

Definition B.1: *Let V be a measure on $[0, \infty)$ such that every finite interval has a finite mass. If the integral on the right side*

$$\hat{V}(s) = \int_{[0,\infty)} e^{-sx} \, dV(x), \tag{B.1}$$

converges for $s > a$ and some $a \in \mathbb{R}$ then $\hat{V}$ is defined for $s > a$ and is called the **Laplace transform** *of V.*

Theorem B.2: *A measure V is uniquely determined by the values of its Laplace transform (B.1) in some interval (a, ∞).*

Note that if V is a probability measure then $\hat{V}(s)$ exists for $s \geq 0$ and $\hat{V}(0) = 1$. If V is a measure such that $\hat{V}$ exists for $s > a$, then $\hat{V}(\cdot + a)$ is the Laplace transform of a measure W, absolutely continuous w.r.t. V and such that $\frac{dW}{dV}(x) = e^{-ax}$ for V-almost every x. If V is such that $\hat{V}$ exists for $s \geq a$, then the function $\frac{\hat{V}(\cdot + a)}{\hat{V}(a)}$ is the Laplace transform of a probability measure W, absolutely continuous w.r.t. V such that $\frac{dW}{dV}(x) = \frac{e^{-ax}}{\hat{V}(a)}$ for V-almost every x.

For a Laplace transform $\hat{V}$ defined on $(0, \infty)$ repeated differentiation implies that $\hat{V}$ is completely monotone, that is, it possesses derivatives $V^{(n)}$ of all orders n and $(-1)^n \hat{V}^{(n)}(s) \geq 0$ for $n \in \mathbb{N}_0$ and $s > 0$. This property is even sufficient:

Theorem B.3: *The following statements hold:*

(i) *A function $\hat{V}$ on $(0, \infty)$ is the Laplace transform of a measure V if and only if it is completely monotone.*

(ii) *If $\hat{V}$ is the Laplace transform of a measure V, then V is a probability measure if and only if the integral (B.1) exists also for $s = 0$ and $\hat{V}(0) = 1$.*

B.2 Mellin transform

For instance in the book of Ditkin and Prudnikov (1965) we find the following

Definition B.4: *Let f be a function defined on $(0, \infty)$ such that there exists $c \in \mathbb{R}$ with $\int_0^\infty x^{c-1}|f(x)|\,dx < \infty$. Then the integral transform*

$$\hat{f}^M(z) \;=\; \int_0^\infty x^{z-1} f(x)\,dx$$

*is called the **Mellin transform** of f.*

The region of convergence of the Mellin transform is a strip in the complex plane paralleling the imaginary axis and containing c. The inverse of a Mellin transform is given by

$$f(x) \;=\; \frac{1}{2\pi i} \int_{c-i\infty}^{c+i\infty} x^{-z} \hat{f}^M(z)\,dz, \qquad x > 0. \tag{B.2}$$

For two functions f and g on $(0, \infty)$ with Mellin transforms $\hat{f}^M$ resp. $\hat{g}^M$ the function $\hat{f}^M \hat{g}^M$ is the Mellin transform of

$$h(x) = \int_0^\infty f\left(\frac{x}{y}\right) \frac{g(y)}{y}\,dy, \qquad x > 0. \tag{B.3}$$

B.3 Stieltjes' moment problem

The moment problem of Stieltjes is the following, compare e.g. Widder (1941): Given a sequence of real numbers $\{\mu_n\}_{n=0}^\infty$ we ask whether or not there exists a measure V on $[0, \infty)$ such that

$$\mu_n \;=\; \int_{[0,\infty)} x^n\,dV(x), \qquad n \in \mathbb{N}_0, \tag{B.4}$$

hence such that the numbers μ_n, $n \in \mathbb{N}_0$, are the moments of V.

Consider the following definitions:

Definition B.5: *The* **moment** $\mu(p)$ *of a polynomial*

$$p(t) = \sum_{k=0}^{n} a_k t^k, \qquad t \in \mathbb{R},$$

with respect to the sequence $\{\mu_n\}_{n=0}^{\infty}$ *is given by*

$$\mu(p) = \sum_{k=0}^{n} a_k \mu_k.$$

Definition B.6: *A sequence is called* **positive** *if the moment of every nonnegative polynomial w. r. t. it is nonnegative.*

The positivity of $\{\mu_n\}_{n=0}^{\infty}$ and $\{\mu_n\}_{n=1}^{\infty}$ is obviously necessary but even more sufficient for a sequence to be a sequence of moments in the sense of (B.4):

Theorem B.7: (cp. Widder (1941)) *For a sequence* $\{\mu_n\}_{n\in\mathbb{N}_0} \subset \mathbb{R}$ *there exists a measure* V *on* $[0, \infty)$ *such that*

$$\mu_n = \int_{[0,\infty)} x^n dV(x), \qquad n \in \mathbb{N}_0,$$

the integrals all converging, if and only if the sequences $\{\mu_n\}_{n=0}^{\infty}$ *and* $\{\mu_n\}_{n=1}^{\infty}$ *are positive.*

Appendix C

Integration formulae

The subsequent propositions are taken from Elstrodt (1996) and Bauer (1990):

Proposition C.1: *A bounded function $f : [a, b] \to \mathbb{R}$, where $a, b \in \mathbb{R}^n$, $a < b$, is Riemann integrable if and only if the points of discontinuity of f on $[a, b]$ form a null set, and then*

$$\int_{[a,b]} f(x)dx = \int_{[a,b]} f(x)d\ell(x).$$

Proposition C.2: *Let $I \subset \mathbb{R}$ be an interval[1] and $f : I \to \mathbb{R}$ Riemann integrable on every compact subinterval of I. Then f is Lebesgue integrable on I if and only if $|f|$ is improperly Riemann integrable on I, and then we have*

$$\int_I f(x)dx = \int_I f(x)d\ell(x).$$

Proposition C.3: (Fubini's theorem) *Let $(X, \mathcal{A}_X, \mu_X)$ and $(Y, \mathcal{A}_Y, \mu_Y)$ be two measure spaces such that μ_X and μ_Y are σ-finite. Then for every $\mathcal{A}_X \otimes \mathcal{A}_Y$-$\bar{\mathcal{B}}$-measurable function $f : X \times Y \to [0, \infty]$ the functions defined by*

$$x \mapsto \int_Y f(x,y)d\mu_Y(y) \qquad resp. \qquad y \mapsto \int_X f(x,y)d\mu_X(x)$$

on X resp. Y are $\mathcal{A}_X$-$\mathcal{B}$-measurable resp. $\mathcal{A}_Y$-$\mathcal{B}$-measurable and we have

$$\int_{X \times Y} f(x,y)d\mu_X \otimes \mu_Y(x,y) = \int_X \left(\int_Y f(x,y)d\mu_Y(y) \right) d\mu_X(x)$$

$$= \int_Y \left(\int_X f(x,y)d\mu_X(x) \right) d\mu_Y(y).$$

For two functions f and g denote by $f \circ g$ the composition of f and g, i.e. $f \circ g(\cdot) = f(g(\cdot))$.

---

[1]Here, intervals include sets like $(0, \infty)$ as well.

Proposition C.4: (Integration w. r. t. induced measures) *Let $(X, \mathcal{A}, \mu)$ resp. $(X', \mathcal{A}')$ be a measure space resp. measurable space. Further let $T : (X, \mathcal{A}) \to (X', \mathcal{A}')$ be an $\mathcal{A}$-$\mathcal{A}'$-measurable mapping and $T(\mu)$ the measure on $(X', \mathcal{A}')$ induced by T. Then, for an $\mathcal{A}'$-$\bar{\mathcal{B}}$-measurable function $f : X' \to [0, \infty]$ we have*

$$\int_{X'} f \, dT(\mu) = \int_X f \circ T \, d\mu.$$

Proposition C.5: (Transformation formula) *Let $X, X' \subset \mathbb{R}^n$ be open, further $h : X \to X'$ a C^1-diffeomorphism with Jacobian $\mathcal{J}_h$. A function $f' : X' \to \bar{\mathbb{R}}$ is ℓ^n-integrable on X' if and only if the function $f' \circ h \cdot |\det \mathcal{J}_h|$ is ℓ^n-integrable on X. Moreover*

$$\int_{X'} f' \, d\ell^n = \int_X f' \circ h \cdot |\det \mathcal{J}_h| d\ell^n.$$

The above Proposition implies the known transformation formula for densities: If especially $h : \mathbb{R}^n \to \mathbb{R}^n$ is injective, and Y_1, Y_2 random variables in $\mathbb{R}^n$ such that Y_1 follows the ℓ^n-density f_{Y_1} and $Y_1 = h(Y_2)$ then Y_2 has the density

$$f_{Y_2} = f_{Y_1} \circ h \cdot |\det \mathcal{J}_h| \tag{C.1}$$

since for $B \in \mathcal{B}^n$

$$P(Y_2 \in B) = P(h(Y_2) \in h(B)) = P(Y_1 \in h(B))$$

$$= \int_{h(B)} f_{Y_1} d\ell^n = \int_B f_{Y_1} \circ h \cdot |\det \mathcal{J}_h| d\ell^n.$$

Proposition C.6: (Differentiation under the integral sign) *Let μ be a measure on $\mathbb{R}$, $I \subset \mathbb{R}$ an interval, $t_0 \in I$, $X \in \mathcal{B}$ and $f : I \times X \to \bar{\mathbb{R}}$ such that*

a) $f(t, \cdot)$ is μ-measurable and integrable for all $t \in I$,

b) the partial derivatives $\frac{\partial}{\partial t} f(t_0, x)$ exist for all $x \in X$, and

c) there exist an integrable function g and a neighborhood $U \subset I$ of t_0 such that

$$\left| \frac{\partial}{\partial t} f(t, x) \right| \leq g(x), \qquad t \in U, \, x \in X.$$

Then, the function $F : I \to \mathbb{R}$, where

$$F(t) = \int_X f(t, x) d\mu(x), \qquad t \in I,$$

is differentiable in t_0 (as the case may be one-sided), $\frac{\partial}{\partial t} f(t_0, \cdot)$ is integrable and we have

$$F'(t_0) = \int_X \frac{\partial}{\partial t} f(t_0, x) d\mu(x).$$

Appendix D

Probability densities and discrete probabilities

Definition D.1: Probability densities on $\mathbb{R}$ w. r. t. the Lebesgue measure:
A random variable X is

(i) **uniformly distributed** *on $[a,b]$ for $a,b \in \mathbb{R}$ with $a < b$, say $X \sim \mathrm{U}[a,b]$, if its density function is given by*

$$f_X(x) = \frac{1}{b-a}\mathbb{1}_{(a,b)}(x) \qquad for \ \ell\text{-almost all } x \in \mathbb{R},$$

(ii) **Gamma distributed** *with parameters $\beta, \gamma > 0$, $X \sim \Gamma(\beta, \gamma)$, if its density equals*

$$f_X(x) = \frac{\beta^\gamma}{\Gamma(\gamma)}x^{\gamma-1}e^{-\beta x}\mathbb{1}_{(0,\infty)}(x) \qquad for \ \ell\text{-almost all } x \in \mathbb{R},$$

(iii) **exponentially distributed** *with parameter $\lambda > 0$, $X \sim \mathrm{Exp}(\lambda)$, if $X \sim \Gamma(\lambda, 1)$,*

(iv) **Beta distributed** *with parameters $a,b > 0$, $X \sim \mathrm{B}(a,b)$, if its density equals*

$$f_X(x) = \frac{1}{\mathrm{Beta}(a,b)}x^{a-1}(1-x)^{b-1}\mathbb{1}_{(0,1)}(x) \qquad for \ \ell\text{-almost all } x \in \mathbb{R}.$$

Evidently, if $X \sim \mathrm{B}(a,b)$ for some $a,b > 0$ we find $1 - X \sim \mathrm{B}(b,a)$.

Definition D.2: *A random variable X with values in $\mathbb{N}_0$ is* **Poisson distributed** *with intensity $\lambda > 0$, say $X \sim \mathrm{Poi}(\lambda)$, if*

$$P(X = k) = \frac{\lambda^k}{k!}e^{-\lambda}, \qquad k \in \mathbb{N}_0.$$

For technical reasons, if $\lambda = 0$ we say $X \sim \mathrm{Poi}(0)$ if $P(X = 0) = 1$.

Simulation of Gamma distributed random variables

Gamma distributed random variables can be simulated using the ratio-of-uniforms method. The following algorithm simulates a $\Gamma(1, \beta)$-distribution with scale parameter 1 and a shape parameter β exceeding 1. It is due to C. H. Cheng and G. M. Feast and can be found e.g. in Gentle (1998):

Algorithm D.3:   *1. Generate u_1 and u_2 independently from $U[0,1]$, and put*

$$v = \frac{\left(\beta - \frac{1}{6\beta}\right) u_1}{(\beta - 1)u_2} \ .$$

2. If

$$\frac{2(u_2 - 1)}{\beta - 1} + v + \frac{1}{v} \leq 2 \qquad or \qquad \frac{2\ln u_2}{\beta - 1} - \ln v + v \leq 1,$$

then deliver $x = (\beta - 1)v$.

3. Go to step 1.

Appendix E

Simulation procedure

```
% This MATLAB program simulates GOS-processes
%
clear;
figure(1);
hold on;
set(gca,'fontsize',18);
xlabel('Time t');
ylabel('State N_t');

aux=2; % switches between examples
SimNum=1; % number of simulated paths
n=4*10^4; % number of simulated occurrence times

q=0.5; % parameter for sequence alpha_i=1-q^i, with 0<q<1

% definition of alpha_i
switch aux
case 1
 alpha=0:n-1; % alpha_i=i-1
 title(strcat('GOS-process w.r.t. 0,1,2,...'));
case 2
 alpha=1-q.^(0:n-1); aux=2;   % alpha_i=1-q^{i-1}
 title(strcat('GOS-process w.r.t. \alpha_i=1-',num2str(q),'^{i-1}'));
case 3
  alpha=zeros(1,n); alpha(2:2:n)=1;  % sequence 01010101...
  title(strcat('GOS-process w.r.t. 0,1,0,1,...'));
end;
% definition of gamma_i
gamma=cumsum(alpha+1);
clear('alpha');
```

```
for i=1:SimNum
  U = rand(1,n-1);
  % transformation into beta-distributed random variables
  % T_i/T_{i+1}~Beta(g_i,1)
  U=U.^(1./gamma(1:n-1)); % vector of beta-distributed r.v.

  % occurrence times T
  T(n)=GammaRandomVariable(gamma(n)); % last occurrence time
  T(n-1:-1:1)=T(n)*cumprod(U(n-1:-1:1));

  % Plot
  stairs(T,0:n-1);
end;
clear('gamma', 'U');
```

The Matlab function GammaRandomVariable(x) implements Algorithm D.3 of Appendix D to simulate a $\Gamma(1,x)$-distributed random variable, $x > 1$:

```
% Simulates a Gamma(1,x)-distributed random variable
% algorithm due to Cheng/Feast
function res=GammaRandomVariable(x)
while (1>0)
u=rand(2);
v=(x-1/(6*x))*u(1)/(x-1)/u(2);
if (2*(u(2)-1)/(x-1)+v+1/v <= 2)
    res=(x-1)*v;
    return;
elseif (2*log(u(2))/(x-1)-log(v)+v <= 1)
       res=(x-1)*v;
       return;
end;
end;
```

Nomenclature

$\mathbb{1}_A$ .... indicator function of a set A, page 5

$\mathcal{A} \otimes \mathcal{B}$ .... product of σ-algebras, page 133

Ai, Bi .... Airy functions, page 125

A^∞ .... $A^\infty = \sum_{i=1}^{\infty} \left(\frac{\alpha_i + 1}{\alpha + 1} - 1 \right)$, page 79

$\mathrm{BesselI}_\nu, \mathrm{BesselK}_\nu$ .... modified Bessel functions of first resp. second kind, page 126

Beta .... Beta function, page 124

$\bar{\mathcal{B}}^n$ .... Borel sets in $\bar{\mathbb{R}}^n$, page 12

$\mathcal{B}^n$ .... Borel sets in $\mathbb{R}^n$, page 5

$\mathcal{B}_+$ .... Borelian σ-algebra on $(0, \infty)$, page 112

$\mathcal{F}$ .... σ-algebra, page 5

F^{-1} .... pseudo-inverse function, page 5

$_p\mathrm{F}_q$ .... Generalized hypergeometric function, page 124

$\mathcal{G}$ .... prob. measures corresponding to GOS-processes, page 94

Γ .... Gamma function, page 123

$\mathrm{G}_{p,q}^{m,n}$, G_n .... Meijer's G-function, page 125

$\mathcal{H}(\mathbb{R}^{[0,\infty)})$ .... σ-algebra on $\mathbb{R}^{[0,\infty)}$, page 7

$\mathcal{H}_t^X$ .... natural history of a stochastic process, page 7

$\mathcal{J}_h$ .... Jacobian of a function h, page 134

K_n, $K_n(t)$ .... ordered subsets of $(0, \infty)^n$, page 5

$\bar{\ell}^n$ .... n-dimensional Lebesgue measure on $\bar{\mathcal{B}}^n$, page 12

ℓ^n .... n-dimensional Lebesgue measure, page 5

μ_n .... n-th moment of a measure, page 130

$\mu(p)$ .... moment of a polynomial p, page 131

$\mathcal{N}$ .... set of counting functions, page 8

$\mathfrak{n}$ .... counting function, page 8

$\mathbb{N}, \mathbb{N}_0$ .... natural numbers excluding resp. including 0, page 125

N .... point process, page 8

$o(h)$ .... Landau symbol, page 9

Ω .... sample space, page 5

$\mathcal{P}$ .... set of probability measures on $\mathcal{N}$, page 93

P .... probability measure, page 5

p .... period of the parametrizing sequence, page 88

$p_{<\infty}, p_\infty$ .... probability that at least one resp. no jump occurs, page 26

Ψ .... iterated integral, page 36

$\bar{\mathbb{R}}, \bar{\mathbb{R}}^n$ .... extended real numbers, $\bar{\mathbb{R}} = \mathbb{R} \cup \{-\infty, \infty\}$, page 12

$\mathbb{R}, \mathbb{R}^n$ .... real numbers, page 5

$\Re(z), \Im(z)$ .... real and imaginary part of a complex number z, page 123

$\mathbb{R}^{[0,\infty)}$ .... real functions, page 7

S_i .... i-th sojourn time of a point process, page 8

$\mathfrak{t}_i$ .... i-th discontinuity of a counting function, page 59

T_i .... i-th occurrence time of a point process, page 8

$T(\mu)$ .... measure induced by T w.r.t. μ, page 5

$U_{i:n}$ .... uniform (generalized) order statistics, page 6

X .... stochastic process, page 7

$X_{i:n}$ .... (generalized) order statistics, page 6

$(x)_n$ .... Pochhammer's symbol, page 124

Bibliography

Abramowitz, M. and Stegun, I. A. (1964). *Handbook of mathematical functions with formulas, graphs and mathematical tables.* U.S. Department of Commerce, Washington.

Bauer, H. (1990). *Maß- und Integrationstheorie.* Walter de Gruyter, Berlin.

Berman, S. M. (1980). Stationarity, isotropy and sphericity in ℓ_p. *Z. Wahrscheinlichkeitstheor. Verw. Geb.*, 54:21–23.

Bronstein, I. and Semendjajew, K. (1996). *Teubner-Taschenbuch der Mathematik. Teil 1.* Teubner, Leipzig.

Burkschat, M., Cramer, E., and Kamps, U. (2003). Dual generalized order statistics. *Metron*, 61(1):13–26.

Cramer, E. (2002). *Contributions to Generalized Order Statistics.* Habilitationsschrift, University of Oldenburg.

Cramer, E. and Kamps, U. (2003). Marginal distributions of sequential and generalized order statistics. *Metrika*, 58(3):293–310.

Crump, K. S. (1975). On point processes having an order statistic structure. *Sankhya, Ser. A*, 37:396–404.

Ditkin, V. A. and Prudnikov, A. P. (1965). *Integral transforms and operational calculus.* Pergamon Press, Oxford.

Elstrodt, J. (1996). *Maß- und Integrationstheorie.* Springer-Lehrbuch. Springer, Berlin.

Erdélyi, A., Magnus, W., Oberhettinger, F., and Tricomi, F. G. (1953). *Higher transcendental functions*, volume I of *Bateman Manuscript Project.* McGraw-Hill Book Co., New York.

Feigin, P. D. (1979). On the characterization of point processes with the order statistic property. *J. Appl. Probab.*, 16:297–304.

Feller, W. (1971). *An introduction to probability theory and its applications*, volume II of *Wiley Series in Probability and Mathematical Statistics*. John Wiley and Sons, Inc., New York, 2nd edition.

Gentle, J. E. (1998). *Random number generation and Monte Carlo methods*. Statistics and Computing. Springer, New York.

Grandell, J. (1976). *Doubly stochastic Poisson processes*, volume 529 of *Lecture Notes in Mathematics*. Springer-Verlag, Berlin.

Grandell, J. (1997). *Mixed Poisson processes*, volume 77 of *Monographs on Statistics and Applied Probability*. Chapman & Hall, London.

Hardy, G. H. (1956). *Divergent Series*. Clarendon Press/Oxford University Press, Oxford, corr. print. of the 1st edition.

Hayakawa, Y. (2000). A new characterisation property of mixed Poisson processes via Berman's theorem. *J. Appl. Probab.*, 37(1):261–268.

Hofmann, M. (1955). Über zusammengesetzte Poisson-Prozesse und ihre Anwendungen in der Unfallversicherung. *Mitt. Verein. schweiz. Versicherungsmath.*, 55:499–575.

Kamps, U. (1995). *A concept of generalized order statistics*. Teubner Skripten zur Mathematischen Stochastik. B. G. Teubner, Stuttgart.

Lundberg, O. (1940). *On Random Processes and their Application to Sickness and Accident Statistics*. PhD thesis, University of Stockholm, Uppsala.

Lundberg, O. (1964). *On random processes and their application to sickness and accident statistics*. Uppsala, Almqvist & Wiksells.

Mathai, A. M. (1993). *A handbook of generalized special functions for statistical and physical sciences*. Oxford Science Publications. Clarendon Press/Oxford University Press, New York.

Nawrotzki, K. (1962). Ein Grenzwertsatz für homogene zufällige Punktfolgen (Verallgemeinerung eines Satzes von A. Rényi). *Math. Nachr.*, 24:201–217.

Niese, B. (2006). A martingale characterization of Pólya-Lundberg processes. *J. Appl. Probab.*, 43(3):741–754.

Paris, R. B. and Kaminski, D. (2001). *Asymptotics and Mellin-Barnes integrals*, volume 85 of *Encyclopedia of Mathematics and its Applications*. Cambridge University Press, Cambridge.

Puri, P. S. (1982). On the characterization of point processes with the order statistic property without the moment condition. *J. Appl. Probab.*, 19:39–51.

Rainville, E. D. (1960). *Special functions.* The Macmillan Co., New York.

Schmidt, K. D. (1996). *Lectures on risk theory.* Teubner Skripten zur Mathematischen Stochastik. B. G. Teubner, Stuttgart.

Schmidt, K. D. and Zocher, M. (2003). Claim Number Processes having the Multinomial Property. *Dresdner Schriften zur Versicherungsmathematik.*

Shaked, M., Spizzichino, F., and Suter, F. (2004). Uniform order statistics property and l_∞-spherical densities. *Probab. Eng. Inf. Sci.*, 18(3):275–297.

Vallée, O. and Soares, M. (2004). *Airy functions and applications to physics.* Imperial College Press, London.

Walhin, J. F. and Paris, J. (1999). Using mixed Poisson processes in connection with Bonus-Malus systems. *ASTIN Bulletin*, 29(1):81–99.

Widder, D. V. (1941). *The Laplace Transform.* Number 6 in Princeton Mathematical Series. Princeton University Press, Princeton, N. J.

Index

Wissenschaftlicher Werdegang

Schule und Studium

1997   Abitur am Gymnasium Dresden-Plauen

1997-2003   Studium der Wirtschaftsmathematik an der Technischen Universität

Dresden und an der Université Joseph Fourier, Grenoble

Abschluss: Diplom

Berufserfahrung

1997-2003   während des Studiums: wissenschaftliche Hilfskraft an der Fakultät

Mathematik und Naturwissenschaften der Technischen Universität Dresden

seit 2003   wissenschaftliche Mitarbeiterin am Fachbereich Mathematik

der Technischen Universität Darmstadt (Arbeitsgruppe Stochastik)